Colliding Galaxies

The Universe in Turmoil

OTHER RECOMMENDED BOOKS BY BARRY PARKER

INVISIBLE MATTER AND THE
FATE OF THE UNIVERSE

CREATION
The Story of the Origin and Evolution of the Universe

SEARCH FOR A SUPERTHEORY
From Atoms to Superstrings

EINSTEIN'S DREAM
The Search for a Unified Theory of the Universe

Colliding Galaxies
The Universe in Turmoil

Barry Parker

Drawings by
Lori Scoffield

Plenum Press • New York and London

Library of Congress Cataloging-in-Publication Data

Parker, Barry R.
    Colliding galaxies : the universe in turmoil / Barry Parker ;
  drawings by Lori Scoffield.
      p.   cm.
    Includes bibliographical references and index.
    ISBN 0-306-43566-7
    1. Active galaxies.   2. Astrophysics.   I. Title.
  QB858.3.P37 1990
  523.1'12--dc20
                                                        90-40651
                                                            CIP

ISBN 0-306-43566-7

© 1990 Barry Parker
Plenum Press is a Division of
Plenum Publishing Corporation
233 Spring Street, New York, N.Y. 10013

Printed in the United States of America

Preface

I remember sitting spellbound, watching the movie *When Worlds Collide*. Two planets hurled through space toward Earth while scientists and engineers frantically raced to complete a rocket-ship that would take them to safety. In the final moments the spaceship lifted off as the occupants watched the Earth bulge, crack, then literally explode as one of the planets struck it.

As I left the theater I wondered if it was really possible for another world to collide with Earth. Later I learned that while many catastrophic collisions no doubt occurred early in the history of the solar system, today they are exceedingly rare. I was relieved, but in another sense I was disappointed (not that I hoped a collision of this type would actually occur). A collision of two objects in space, say, two stars, I was sure would be a spectacular event.

It is quite unlikely, however, that we will ever witness the collision of two stars. The event is just too rare. But collisions of systems of stars—galaxies—oddly enough, are relatively common. In fact, we see evidence of several in the sky right now.

In this book I will consider both colliding and exploding galaxies. I will begin with a brief introduction to ordinary galaxies, then turn to exploding galaxies. From there I will go to colliding galaxies and show how astronomers have been able to simulate them using computers. Finally, I will turn to the overall universe and the forces that have shaped it.

It's a fascinating story, a story of discovery and, as we shall

v

see, a story that contains controversy. But perhaps most of all it's a story of people. I have tried to give insights into the lives of some of the scientists who have worked on the problems, and their struggle to understand galaxies and the universe. Hundreds of scientists have contributed to our present understanding, and it would be impossible in a book such as this one to include all of their contributions. I apologize for those I have omitted.

It is also difficult in a book of this type to get around the use of technical terms. I have avoided them as much as possible, but for those unfamiliar with some of the words I use, I have provided a glossary at the end of the book. Very large and very small numbers are also a problem. To write them out in detail would be cumbersome; I have therefore used scientific notation. In this notation a number such as $1/10,000$ is written as 10^{-4}.

I am particularly grateful to the scientists who assisted me. Interviews were conducted, in most cases by telephone or letter, with many of the people mentioned in the book. In many cases they also supplied photographs and reprints. I would like to express my appreciation to them. They are Halton Arp, Joshua Barnes, Mitchell Begelman, Kirk Borne, Jack Burns, Eric Feigelson, Richard Green, James Gunn, Paul Hodge, John Huchra, William Keel, Kwok-Yung Lo, Jeremiah Ostriker, Richard Perley, Sterl Phinney, George Rieke, Paul Schecter, Ethan Schreier, François Schweizer, Alar Toomre, Daniel Weedman, Simon White, and Susan Wyckoff.

The sketches, paintings, and line drawings were done by Lori Scoffield. I would like to thank her for an excellent job. I would also like to thank Linda Greenspan Regan and the staff of Plenum for their assistance in bringing the book to its final form. And finally, I would like to thank my wife for her support while the book was being written.

Contents

CHAPTER 1

Introduction

Looking upward on a clear summer night in the country you see a breathtaking panorama. Stars dot the sky from horizon to horizon. And almost directly overhead a silvery band of light runs from the northern to the southern horizon—the Milky Way. Like a misty stream it meanders through the velvet darkness, caressed here and there by patterns of stars.

Toward the south it gets wider and denser, then suddenly it is interrupted by a dark band—the Great Rift. Beyond the Rift it continues bright, then gradually dims as it approaches the horizon. If you take a pair of binoculars and scan it, you see that it is composed of millions of stars—actually billions. This is the system—the island universe of stars—that we live in. We call it the Milky Way galaxy.

Scanning further, you see that it is composed not just of stars. Glowing patches of gas are faintly visible, and here and there you see dark streams. They look like channels through the stars, but are actually streams of dark gas and dust that cut off our view of the stars behind them.

It's a stunning sight, but there is much more that cannot be seen with binoculars. Planets, most of them dead worlds pocked with craters, orbit many of the stars. A few, however, likely contain lush forests, lakes, and streams as ours does. Tiny dense stars are also present, some the size of the Earth, others only a few miles across. Some shine steadily with a dim glow, others are like tiny lighthouses with searchlights shining out into the

The Rosette Nebula. A gaseous cloud within our galaxy. (Lick Observatory, University of California, Santa Cruz, Calif. 95064)

darkness. And there are bizarre objects—black holes—detectable only through their strong gravitational field. Anyone venturing too close to one would drop into a bottomless pit.

Our galaxy is so large that it takes a beam of light (traveling at 186,000 miles per second) 100,000 years to cross it. Yet beyond it are hundreds of billions of other galaxies; they stretch as far as telescopes can reach. We see no end to them. Equally amazing is the fact that no two of them are exactly the same. They are as identifiable as people, but just as people of different nations

A spiral galaxy. (Mount Wilson Observatory and Palomar Observatory)

have different overall characteristics, so too do galaxies. We can, in fact, group them into three major classifications: spirals, ellipticals, and irregulars. Spirals, as their name suggests, are disk-shaped with long spiral arms. Ellipticals are more dense and generally egg-shaped, while irregulars are nonsymmetric and have no regular form.

If you take a moment to think about them, galaxies are truly

amazing objects. In the view of a large telescope you can sometimes see hundreds, each containing billions of stars. I remember staring at photographs of them when I was young, wondering what it would be like to travel to one. What would we learn?
Were there any advanced civilizations within them? I wanted to
learn more about them, and, in particular, to see them through
large telescopes. So, near the end of my senior year at university, I applied to a nearby observatory for a summer job.

When the school year ended I still hadn't heard anything,
so I phoned the office that handled the applications. "Sorry, we
can't do anything over the phone, you'll have to come down to
the office," I was told.

So down I went. After introducing myself to the man behind the counter I asked if the summer jobs at the observatory
had been filled. He checked through some pages. "Yes, everyone has been selected," he said.

My heart thumped. "Is my name among them?" I asked.

"I'll have to get the application forms," he said as he turned
and went to another room. A few moments later he returned
with a pile so high it made my mouth drop.

"Did that many people apply?" I asked in disbelief.

He nodded as he began thumbing through the top ones.
"No, your name is not among those selected," he said.

I felt my body go numb. He then began thumbing down
through the pile looking for my application. He found it, pulled
it out and looked at it for a few moments, then placed it on the
top of the pile.

"Sorry," he said.

I left, still in a bit of a daze, and wondering where I would
get another job.

Two days later I received a telephone call. I recognized the
voice immediately; it was the man from the employment office.
"Are you still interested in the job at the observatory?" he asked.
"One of the other people turned it down."

I could hardly contain my joy. "Yes, I'm still interested," I
said. And a few weeks later I had my first view of a galaxy
through a large telescope. It was a magnificant sight.

Section of a gaseous nebula within our galaxy. (Lick Observatory, University of California, Santa Cruz, Calif. 95064)

What I didn't realize at the time, though, is how diverse and complex galaxies are. We now know that many galaxies exhibit evidence of tremendous violence at their core. Filaments of gas are seen protruding tens of thousands of light-years (the distance light travels in one year) into space. In other cases the gravitational interaction between galaxies hurls gas and stars far into intergalactic space, creating huge arching streams. In a few cases a more violent collision causes the core to blaze with the light of millions of new stars.

Most of the energy that is released comes out in the form of radio waves or X rays. The tremendous violence that is occurring is, in most cases, invisible when viewed through optical telescopes. It has therefore been radio telescopes (and later, X-ray telescopes) that have allowed us to make most of the major breakthroughs in this area.

Interestingly, the first instrument that detected radio waves from space was not built to look for them. It was built to pinpoint the source of an annoying hiss in long-distance telephone calls. Customers at Bell Labs were beginning to complain—they could hardly hear anything at times because of the hiss. Bell Labs therefore assigned Karl Jansky to find out what was causing it. Jansky built a large, unwieldy antenna, and within a short time he had detected several possible sources. Among them was a strange signal that was eventually traced to space. Jansky later showed that it was coming from the center of our galaxy. It was an exciting discovery, and Jansky tried for years to get astronomers interested in it. But it was far out of the mainstream of astronomy, and few astronomers took an interest.

His articles did, however, catch the attention of an electrical engineer in Wheaton, Illinois, by the name of Grote Reber. Reber built an antenna—a dish that, interestingly, is very similar to the radio telescopes we use today. With it he was able to detect radio waves. And for almost a decade he was the one and only radio astronomer in the world.

Although radio astronomy started slowly, it blossomed quickly after the war. The breakthrough was the development of radar. Radar was used extensively in England throughout the war to detect enemy planes. Many people were trained in it, and when the war ended most of them went back to university looking for research projects related to radar.

Several important scientific discoveries related to radar had, in fact, been made during the war. The sun was shown to be a strong source of radio waves. Hydrogen gas out in space was also shown to emit radio waves. And, of course, there was Reber's discovery in the United States that the center of the Milky Way was a source of radio waves.

The time was ripe. Scientists soon began building antennas to confirm and extend these discoveries. Research groups were organized at Cambridge University, Manchester University, and other universities. Similar interests soon developed in Australia, Holland, and the United States.

But the sky was "fuzzy" through a radio telescope. They didn't have the sharp resolution (ability to distinguish two closely spaced objects) of optical telescopes, and astronomers had difficulty associating the radio sources with known optical sources. Eventually, though, the problem was overcome and the first sources were identified. They were stars—exploding stars. Then a radio galaxy was identified in the constellation Cygnus. When astronomers finally located it with optical telescopes they were amazed. It appeared to be a double galaxy, or perhaps two galaxies in collision. Considerable excitement was generated; then calculations showed that such collisions were possible, and if they did occur the object would become a strong source of radio waves.

Was this our first view of the collision of two galaxies? It seemed possible, but, as radio telescopes increased in size, astronomers noticed that the radio emission wasn't coming from the galaxies themselves (assuming there were two), but from two diffuse patches, or lobes, on either side of them. Furthermore, they were millions of light-years away from the galaxies. The huge clouds of radio-emitting gas were, in fact, much larger than the galaxies.

What was the connection between the galaxies and the lobes? Were the lobes being created by the collision? It seemed unlikely. Soon similar lobes were found near other galaxies— galaxies that didn't appear to be colliding. Perhaps the object in Cygnus wasn't a collision of two galaxies after all. Soon there

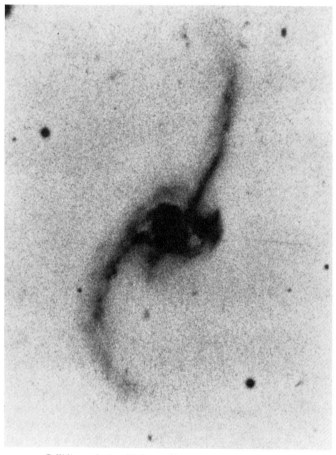

Colliding galaxies. (Palomar Observatory and Halton Arp)

was no doubt. Looking at the object more closely, astronomers realized it was a single galaxy with a dark band of dust through its center.

As astronomers continued to study the lobes they found they were incredibly energetic. They had an energy equivalent to the annihilation of millions of stars. Then regions of high temperature—"hot spots"—were discovered on many of the lobes—usually on the side facing away from the galaxy. And finally, what was perhaps the key discovery: Astronomers detected jets, or narrow beams of hot gas emerging from the core of the galaxy—leading to the lobes. It seemed as if the hot gas was being ejected at incredible speeds into a tiny funnel. At the ends of the funnel it was spreading out, creating the lobes.

But what was powering the system? Whatever it was, it was producing an incredible amount of energy, and it was deeply embedded within the nucleus of the galaxy. Interest began to center around the idea that it might be a black hole. A black hole seemed to be the only thing that could produce energy on such a huge scale. But it couldn't be an ordinary black hole; it had to be an excessively massive one. It had to have a mass millions of times greater than our sun.

Astronomers had barely gotten used to the tremendous energies associated with these galaxies when, in the early 1960s, a discovery was made that completely overshadowed them. Objects (now called quasars) were discovered in the outer regions of the universe that had energy outputs millions of times greater. What were they? Ideas poured out in the journals, but confusion persisted for years. No one was sure what they were, or what was causing their energy. Scientists weren't even sure for years that they were really located in the outer regions of the universe. The controversy continued, but gradually proof came showing that they were indeed distant objects. Then, finally, a number of important breakthroughs were made, and soon quasars were much better understood.

After the galaxy in Cygnus was shown to be an active galaxy (a galaxy with an exploding core) rather than the collision of two galaxies, interest in colliding galaxies declined. Nevertheless, there were objects in space that appeared to be interacting. In a number of cases long luminous filaments could be seen arching out from galaxies. In other cases luminous bridges linked pairs of galaxies. For lack of a better word astronomers called them peculiar galaxies.

What was causing these filaments and bridges? Fritz Zwicky

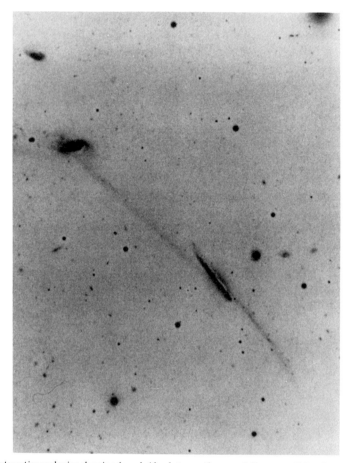

Interacting galaxies showing long bridge between them, and filaments. Printed as negative to show filaments more clearly. (Palomar Observatory and Halton Arp)

of Mount Wilson Observatory began studying them. He soon became convinced that they were due to gravitational interactions. About the same time, however, Boris Vorontsov-Velyaminov of the USSR made a similar study (he also assembled an atlas of them). He took a different point of view, suggesting that they were due to magnetic fields.

The 200-inch reflector and dome at Palomar. (Palomar Observatory)

Then, in 1966, Halton Arp of Mount Wilson and Palomar Observatories published another atlas. His atlas had a much wider distribution than Vorontsov-Velyaminov's, and the photographs were much clearer. Astronomers soon became intrigued with the objects in it. A few began trying to duplicate the filaments and arms in computer simulations of collisions.

Some of the first simulations of this type were done by Alar and Juri Toomre of New York University. Using computers at the Goddard Institute for Space Studies, they showed that grav-

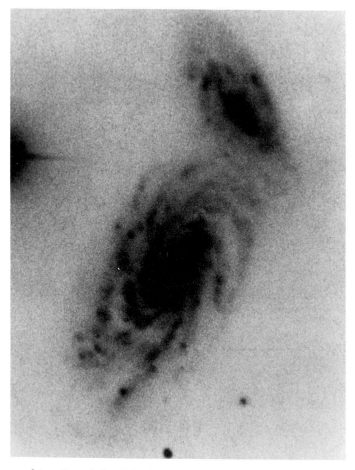

Interacting galaxies. Printed as negative to show filaments more clearly.

itational forces were important, and that filaments and bridges would be drawn out from galaxies as they passed one another in space.

In this book I will talk about both colliding and active (exploding) galaxies. You will see many photographs of exploding and colliding galaxies, and also photographs of computer simulations of collisions and interactions. You will no doubt be amazed at how close some of the simulations are to the real thing. In looking at the photographs you will see that galaxies do not have to physically collide to interact. If they pass close enough to one another they will interact, and long filaments and arms will be drawn out. In a direct collision, filaments may also be drawn out, but, in general, much more damage will be done; the galaxies may even merge. As we will see, however, none of the stars themselves actually collide.

Astronomers can now simulate both passing interactions and direct collisions and mergers. Alar and Juri Toomre's work was just the first step. Over the years computers have become larger and more powerful, and as a result simulations have become much more accurate. The Toomres were able to use only a few hundred points to represent a galaxy. Today astronomers use tens of thousands.

What do simulations tell us about the merger of two galaxies? Unfortunately, this question has not been fully answered. We will see, in fact, that it has generated considerable controversy in the world of astronomy. Some astronomers believe that the collision of two disk (spiral) galaxies produces an elliptical galaxy. Indeed, a number of astronomers go as far as saying that all ellipticals were at one time produced in collisions of disk galaxies. Many others, however, do not accept this. We will look at this controversy in detail.

We will also see that our galaxy is of considerable interest. At first glance it seems docile, yet we now know that something strange is going on. The core of our galaxy is much more energetic than most people think. It might even be exploding. Long tendrils have recently been discovered emanating from it. Furthermore, our galaxy also seems to be interacting with a nearby pair of galaxies called the Magellanic Clouds. A large stream of gas stretches between the systems.

In the last chapters of the book we will examine the large-scale structure of the universe. We will see that galaxies have a tendency to cluster, and that even clusters tend to group themselves into what we call superclusters. Between these superclusters are huge voids—regions where there are no galaxies. Overall, the universe has a strange mottled appearance—like a sponge, or perhaps like soapsuds. (The voids do, in fact, look like gigantic bubbles.) It is evident that this structure was generated as a result of considerable turmoil in the early universe.

But what caused the turmoil? Astronomers have ideas, but many problems remain. The large-scale structure of the universe, and the problems associated with how it was generated, are at the frontiers of cosmological research. Much has been learned in the last few years, and important breakthroughs will no doubt be made in the near future.

To understand these advances we must start at the beginning. I will therefore begin in the next chapter with the discovery and early study of galaxies.

CHAPTER 2

Galaxies

The mystery of the white nebulae began soon after astronomers first saw them in the sky with their telescopes in the mid-1700s. What were these strange white patches? The German philosopher Immanuel Kant suggested that they might be "island universes" of stars—systems like our own Milky Way consisting of millions of stars. He believed they were so far away that individual stars could not be distinguished. But not everyone agreed with him. Others thought they were huge clouds of gas that were relatively nearby, and a few even thought that they might be the first stages of planetary systems. Many years would pass, though, before the argument would be resolved. Telescopes were still too crude at this stage to provide a solution to the mystery.

Though astronomers could not determine what the white nebulae were, they could catalog them. And in the late 1700s two catalogs were published. The first was the work of the French comet hunter Charles Messier. Messier cataloged them, not because he was interested in them, but because he considered them nuisances. He was a comet hunter, and, to him, comets were important; he had no interest in white nebulae. But white nebulae, and other similar objects, were frequently mistaken for comets. Messier therefore decided to make up a table of these fuzzy nuisances so that comet hunters could avoid them. He published his table in 1784. It contained 103 objects (each is now referred to with an M followed by a number, e.g., M31, M81).

A much more extensive catalog was published a few years later. William Herschel, a German musician turned astronomer, who had emigrated to England in 1757, decided to extend Messier's catalog. He was certain that Messier had missed a few of these fuzzy white objects. By 1786 he had found another 1000. He presented his catalog to the Royal Society of London; then three years later he presented an addition to it that contained another 1000. In 1802 he added a final 500.

When Herschel died in 1822 his son John traveled to South Africa and extended the catalog to the southern hemisphere. By 1864 it contained over 5000 entries. But even with such a large number of objects it is, strangely, not the catalog that astronomers use today. A few years after John Herschel's death the American astronomer J. L. Dreyer began another catalog. When completed, it, along with its supplements, contained 15,000 objects. It is now referred to as the New General Catalog (NGC) and objects in it are labeled with the letters NGC followed by a number (e.g., NGC 587).

THE LEVIATHAN OF PARSONSTOWN

Although thousands of white nebulae had been cataloged by the mid-1800s, little progress had been made in understanding them. William and John Herschel studied them in detail, but had no idea what they were. It soon became obvious that if astronomers were to solve the mystery of the white nebulae, larger telescopes were needed. William Parsons of Ireland began to realize this about 1825. Born in 1800, Parsons became the third Earl of Rosse in 1841. Finding that he had considerable free time, he decided to use it to build telescopes. His first efforts were a series of trial and error, with many failures. Finally, however, he successfully cast a 36-inch mirror. Bolstered by this success, he decided to cast a much larger one, and in the early

Spiral galaxy in the constellation Virgo. (Hale Observatories)

1940s he started work on a mirror of 72-inch diameter. It was a formidable undertaking, as 72 inches was considerably larger than Herschel's largest. But Parsons was confident, and over a period of about four years he was able to cast the mirror and build a housing structure for the telescope. When assembled, it was an impressive sight (the tube was 54 feet long) and people from all over Europe came to see it. It became known as the "Leviathan of Parsonstown." And for over 75 years it was the world's largest telescope.

With the telescope completed, Parsons began going through Herschel's catalog, looking at each of the objects in turn. One that he found to be of particular interest was M51. Looking carefully Parsons saw that it had a strange spiral structure. He sketched it, then began looking for other objects with a similar structure. He found several. When he presented his findings to the Royal Society of London they caused a sensation. What were these strange spirals? Parsons carefully measured their sizes, noted their brightness, and sketched the knots along the arcs. He was convinced they were systems of stars, so far away that individual stars could not be resolved. But he knew there was considerable controversy about them, so he kept his views to himself.

Soon after Parsons' death a new era in the history of astronomy began. Until then the only permanent records of objects in the sky were sketches. But photography was now coming into its own and would soon replace sketching. Equally important was another discovery that was developed about this time: spectroscopy. It had been known since Newton's time that when light passed through a prism it separated into a spectrum of colors. Later it was discovered that the light from a glowing gas, when passed through a slit and prism, exhibited bright-colored lines (now called spectral lines) that were characteristic of the gas. And finally, in 1902, William Wollaston discovered dark lines superimposed on a bright continuous spectrum. They were later studied in detail by the German optician Joseph Fraunhofer.

Spectroscopy (the study of spectral lines) was developed by Gustav Kirchhoff and Robert Bunsen of Germany during the mid-1800s. But it was not applied to astronomy until 1859 when William Huggins, a wealthy English amateur, built a spectroscope and attached it to an 8-inch telescope in London. Huggins, who had been trained in chemistry, heard of Kirchhoff's discoveries and decided to use them in astronomy. He began looking at the spectra of stars, noting that they consisted of dark lines superimposed on a bright background. After studying the spectra of stars for about a year, he turned his spectroscope to the nebulae. And what he saw amazed him. Many of them exhibited bright lines; others, however, gave a spectrum similar to that of stars. What did this mean? To Huggins it meant that there were two types of nebulae. The ones that gave bright line spectra had to be gaseous, since heated gases here on Earth gave bright lines. Similarly, those that exhibited spectra like stars had to be starlike.

It was an important breakthrough, but Huggins was not finished. In 1866 he heard of a discovery made by the Dutch physicist Christian Doppler. Doppler showed that there was a change in frequency (pitch) when a source of waves such as sound moved either toward or away from an observer. This became known as the Doppler effect. A few years later Armand Fizeau showed that it also applied to light. Huggins decided to see if he could measure the effect in the case of stars. He began with the bright star Sirius. There was, indeed, a slight shift of its spectral lines compared to the positions they have when the source is at rest. He calculated the associated speed: It was 30 miles per second in a direction away from Earth. Huggins quickly made measurements on the spectra of other stars, and indeed, most of them had motion relative to Earth. A new tool was available. Astronomers could now determine the velocity of an object toward or away from Earth merely by observing its spectral lines.

The keys to a deeper understanding of the white nebulae had been developed. With spectroscopy, the Doppler effect, and the use of photography, tremendous progress would soon be made.

THE GREAT DEBATE

But if astronomers wanted to learn more about the white nebulae they had to learn more about our own galaxy, the Milky

Way. Herschel had determined that it was shaped like a pancake with a diameter of about 16,000 light-years. His son John concluded, after extensive study, that it was disk-shaped, and when Parsons discovered spirals, astronomers began to wonder if it wasn't, perhaps, a spiral.

In the early 1900s the Dutch astronomer Jacobus Kapteyn organized a worldwide effort to determine the Milky Way's shape and size, and the sun's position in it. He found that it was a spiral with a diameter of about 25,000 light-years. He also determined that the sun was at its center.

But did the white nebulae lie inside or outside the Milky Way? If inside it, they had to be considerably smaller, but if they were on the outside they could be about the same size. In the early 1900s opinion was divided.

One of the central figures in the debate was Harlow Shapley. Born in Missouri in 1885, Shapley began his career as a crime reporter. He covered everything from a gun duel that took place in front of a newspaper office to exposing a charlatan with a counting horse at a nearby circus. But he soon became dissatisfied; it seemed he was being continually passed over when it came to important assignments. What he needed, he told his younger brother, "was to get educated."

So, after completing high school, he headed for the journalism department at Missouri University. When he got there, though, he found to his dismay that it wasn't going to open until the following year. "There I was, all dressed up for a university education and nowhere to go," he later wrote. Determined not to lose a year he selected astronomy and "from there on things went swimmingly. I had found my field," he said. He worked with the small telescopes of the university and taught elementary astronomy classes. And upon graduation he won a fellowship to Princeton University.

At Princeton he worked under the renowned astronomer Henry Norris Russell. Russell was quiet and reserved, and as a result relations between the two men were somewhat strained during the first few days they were together. At the time Russell was working on double (binary) star systems, and one of his major problems was determining the orbits of the two stars. He explained his work to Shapley and a few days later Shapley came to him with the solution to one of the problems. Russell was so amazed that his attitude toward Shapley quickly changed. They soon became close friends. Shapley continued working on binary systems and soon became an expert in the area. Together he and Russell developed many of the techniques now used on such systems, and they published several papers together. "Those were happy days for both of us," Shapley said later.

Shapley's thesis was published in the prestigious *Astrophysical Journal*. And by the time he graduated, his work had come to the attention of astronomers around the world. Hale, the director of Mount Wilson Observatory, noticed it and offered him a job.

Shapley arrived at Mount Wilson in 1914 and was soon working with the 60-inch telescope. For the first few months he spent most of his time assisting others and learning the routine, but he soon developed his own observing program. While on his way to Mount Wilson he stopped off at Harvard Observatory. As he was leaving the director, Solon Bailey, took him aside and gave him some advice. "Study the stars in globular clusters [spherical groups of several hundred thousand stars]," he said. Shapley thought about the suggestion and when he began his own observing program he decided to act on it.

Many of the stars in globular clusters, Shapley soon found, were cepheid variables—stars that varied periodically in light intensity. Cepheids came to the attention of astronomers in 1912 when Henrietta Leavitt of Harvard College Observatory noticed large numbers of them in two irregular galaxies in the south called the Magellanic Clouds. She showed that the luminosity, or absolute brightness, of these variables was related to their period. Eljnar Hertzsprung later used this relation to determine the distance to the Magellanic Clouds. He found them to be 30,000 light-years away. Few took notice of his result at the time, however, because it was accidentally published as 3000 light-years. We now know that they are 200,000 light-years away.

Shapley realized that he could use the Leavitt period-luminosity relation to determine the distance to globular clusters. After considerable work he determined the distance of about 100 of them. Plotting his results on a graph, he saw that the majority of these clusters were to one side of us. The most reasonable explanation of this, he felt, was that they were distributed uniformly around our galaxy and we were off to one side of it. But his measurements also allowed him to determine the size of our galaxy. He got a diameter of 300,000 light-years—about 10 times the accepted value at the time. Most astronomers were able to accept the fact that we weren't at the center of our galaxy, but few would accept the incredible increase in size.

The major problem with the large size was that it meant that many of the white nebulae were inside the system. A few could be outside, but they would be much smaller than the Milky Way. Shapley had, at one time, believed that the white nebulae were independent systems like the Milky Way, but after his work on globular clusters he changed his mind. He was further convinced by some measurements made by a fellow astronomer and close friend, Adrian van Maanen. Van Maanen used a blink comparator—a device that allowed him to blink back and forth between photographs of nebulae taken several years apart—in an attempt to measure the rotational speed of several nebulae. His measurements indicated that there was a slight shift in the positions of the stars. This meant that the nebulae had to be relatively close. If they were far away, shifts of this type could not be measured. It was an extremely difficult measurement but van Maanen was convinced that he had made no errors. And apparently he was able to convinced Shapley as well.

But not everyone shared Shapley's view. Considerable evidence existed indicating that the white nebulae were well beyond our system—independent systems just as large as ours. The first of this evidence came in 1917 when George Richey of

Mount Wilson was looking over some photographs he had taken. To his surprise he saw a tiny point of light in nebula NGC 6946 that had not been there in previous photographs. It had to be a nova, or exploding star, he told himself. He quickly looked through the other plates of nebulae that had been taken at Mount Wilson. And—sure enough—he found several others. News of the discovery soon leaked to Lick Observatory, also in California, and Lick observer Heber Curtis decided to look through the Lick plates. Like Richey, he also found several novae. A simple calculation told him that if they were similar to the novae within the Milky Way, the nebulae had to be extremely distant. In fact, they had to be independent island universes of stars.

But there was a problem. All of the novae did not give the same result. Two appeared to be different from the others. One had taken place in the Andromeda nebula in 1885 and one in a nebula in Centaurus in 1895. They were considerably brighter than the others. Curtis didn't know what to make of them so he ignored them.

The evidence, therefore, was conflicting. If the nebulae were out where Curtis said they were, Shapley showed, using van Maanen's measurements, that their outer regions had to be rotating at a speed faster than that of light. But they couldn't rotate that fast; the speed of light is the uppermost speed allowed in the universe. And what about the two exceedingly bright novae? Both became so bright they rivaled the intensity of the nebula itself. If this was a system of millions of stars, or perhaps even billions, there appeared to be no way that a single exploding star could become as bright as the entire system.

Curtis became quite critical of Shapley's work and Shapley was equally vocal in pointing out that Curtis' evidence was contradictory. It was perhaps inevitable that they would clash. The confrontation occurred in April, 1920 at the National Academy of Science meeting in Washington, D.C. Shapley and Curtis were each asked to give a short talk and a rebuttal. Shapley agreed immediately; Curtis was hesitant at first, but finally agreed on the condition that they be allowed to challenge each other's views. Without this, he felt, the talk would be too dull. Also, he wanted to talk primarily about spiral nebulae. Shapley agreed to the challenge, but wanted the talk to be mostly about our galaxy. A compromise was finally reached.

There were about 200 to 300 people at the debate; among them was Einstein. And true to form, the two men stuck to what they knew best. Shapley talked about the size of our galaxy, our position in it, and his work on globular clusters. Only at the end of the talk did he mention that the size of our galaxy prohibited the spiral nebulae from being systems such as ours. They had to be small satellite systems.

Curtis, on the other hand, talked mainly about the discovery of novae in a number of spirals. He described calculations showing that if they were similar to the novae in our galaxy, the spirals had to be independent systems beyond our own. He criticized Shapley's use of cepheids as distance indicators, and expressed doubt as to the accuracy of van Maanen's results.

Interestingly, the talks were not only, for the most part, on different subjects, but they were also given at different levels. The men had been instructed to make their talk intelligible to a general audience. Shapley did, but Curtis used technical terms freely and went into considerable detail about his methods.

Who won the debate? Certainly, at the time nobody won. In retrospect both men were right in some of their arguments and wrong in others. Shapley's model of our galaxy was correct, though it was too large. Curtis was convinced that we were at the center and he was wrong in this. But Curtis was right in his contention that the spiral nebulae were galaxies, independent of our system.

THE DEBATE IS OVER

The debate did not resolve the controversy, but it did bring the problem to astronomers' attention, and it clarified many of the issues. One of the problems was related to photographs of the nebulae. Although stars had been resolved in the outer regions of some of them, astronomers refrained from calling them stars. They were called "nebulous stars"—mostly because many astronomers were still not convinced at this stage that they were, indeed, stars.

The resolution of the debate finally came from the work of Edwin Hubble. Hubble arrived at Mount Wilson late in the summer of 1919. He had worked on white nebulae while doing his Ph.D. thesis at the University of Chicago. During the course of his thesis research he had become convinced that spirals were separate systems of stars at some distance from our system. With the large telescopes he would have at his disposal at Mount Wilson he was sure he would be able to prove it.

Born in 1889 in Missouri, Hubble was the fifth of seven children. By the time he was 12 he had already developed an interest in astronomy. Shortly after his twelfth birthday he got a letter from his grandfather asking several questions about the planet Mars. His grandfather was so delighted with the reply that he had it published in the Springfield paper. You might say this was his first published paper in astronomy.

At school he was the star of the football team and captain of the track team. In track, the high jump was one of his favorite events. Shortly after he was married, his wife Grace asked him, "How tall do you think I am?"

"You're exactly 5 feet, 4 inches," he replied. She looked surprised. "How did you know?"

He smiled. "I was a high jumper, my dear." And interestingly, he won the high jump in his senior year with a jump of exactly 5 feet, 4 inches.

Upon graduation from high school he received a scholarship to the University of Chicago where he worked in the laboratory of Nobel Laureate Robert Millikan. While he was at the University of Chicago he won a Rhodes scholarship and used it to travel to England where he studied law at Oxford. Upon returning to the United States he practiced law for a short time

Hubble as student at Oxford.

but soon became dissatisfied. After thinking it over he realized astronomy was what he was really interested in, so he closed his law office and went back to university.

He got his Ph.D. from the University of Chicago where he worked with the telescopes at Yerkes Observatory. Although he had been heavily involved with athletics earlier at the University of Chicago, and at Oxford, he no longer had time for them now. To stay in shape, however, he did occasionally swim in nearby Lake Geneva. One day he was walking along a pier on the lake when the wife of a middle-aged college professor fell in. Hubble raced up to where she had fallen and dove in after her. As he tried to get hold of her she struggled so hard he realized it was going to be impossible to rescue her unless he knocked her out. But he was reluctant to do so. Checking the bottom, he discovered that if he placed her on his shoulders her head was above water. Doing this he walked until he got her to shallow water.

He was surprised by her husband's reaction when he delivered her to him. His "Thank you" was cold, Hubble said, and he didn't seem particularly glad to see her.

As he was finishing his thesis, Dr. Hale of Mount Wilson came to Yerkes and offered him a job. But within days World War II broke out and he enlisted.

After the war Hubble returned to the United States. Within a short time he was at Mount Wilson working on spiral nebulae. He took long exposures of several nearby nebulae, then studied them carefully. The mottled lumps within them looked like stars—but were they stars? He took more photographs. Finally, he convinced himself that they were, indeed, stars.

Several astronomers had discovered novae in these nebulae so Hubble kept his eyes open for them. Within a short time he found several. Whenever he discovered one he would put an N next to it to designate it as a nova. One day in 1923 he was checking on a nova in the Andromeda nebula (M31), comparing it on several different plates, when he discovered it was varying periodically in brightness. This meant that it wasn't a nova, after all; it was a variable. He therefore wrote VAR next to it. He hoped it would be a cepheid variable, for if it was he would be able to use the technique Shapley had used earlier on globular clusters to obtain the distance to the Andromeda galaxy. Within a short time he showed that it was, indeed, a cepheid, and using the cepheid period-luminosity relation, he determined the distance to M31 to be approximately 800,000 light-years. (It is now believed to be about 2 million light-years away.)

He wrote to Shapley telling him of the discovery, but Shapley apparently didn't take it seriously. He did, however, mention it in passing to Russell. And Russell did take it seriously.

Hubble continued looking for cepheids and soon found several more in the Andromeda galaxy and in other nearby galaxies. In each case he got distances well beyond the edge of our galaxy. In late 1924 Russell wrote to him offering to announce the results at the upcoming meeting of the American Association for the Advancement of Science in Washington, D.C. He informed him that there was a prize of $1000 for the best paper and he would have an excellent chance of winning it. Hubble agreed to send him a paper.

On the evening before the meeting Russell had dinner with another astronomer, Joel Stebbins, in Washington. He asked Stebbins if he had received Hubble's paper. Stebbins said that he hadn't. "Well, he is an ass," said Russell. "With a perfectly good $1000 available, he refuses to take it." They talked about the problem, deciding that there was still time if Hubble could rush something to them. They decided to telegraph him, and went to the hotel lobby to draft a telegram. As they left for the telegraph office they passed the hotel desk. Russell noticed a large brown envelope on the floor behind the clerk. Enquiring about it he found it was for him—it was Hubble's paper.

Russell read the paper and Hubble won the prize (or at least shared it) for the best paper. Both Shapley and Curtis were at the meeting. "The announcement was dramatic," Stebbins wrote later. "The entire society knew that the debate had come to an end." Using cepheids, Hubble had determined the distance to M31 (the Andromeda nebula), M33, and NGC 6822. They were all well beyond the bounds of the Milky Way, which proved they were indeed island universes of stars.

But there were a few diehards. There was still van Maanen's measurements to explain. They indicated that spirals were relatively nearby. Van Maanen had a reputation for accuracy in making difficult measurements of this type, and it seemed unlikely that he could have made an error.

The situation became even more cloudy when the Swedish astronomer Lundmark visited Mount Wilson in 1923. He made the same measurements and got only about one-tenth the shift van Maanen did. No one was sure who was correct, but in time interest in van Maanen's measurements began to fade. The final blow came in 1935 when Hubble decided to make the measurements himself. He got results that were even less than Lundmark's. Van Maanen had obviously made an error, but even today we are still not certain what his error was.

THE CLASSIFICATION OF GALAXIES

Well before the debate on spirals was over, Hubble had begun working on another problem. He wanted to get a consistent, but simple classification of galaxies that would include all types. Several classification schemes had been introduced over the years. William Herschel published one in which he merely classified them as bright or faint, large or small. His son John extended the system, but it was cumbersome and not of much value. A simple description was needed, and in 1922 Hubble began looking for such a system. The incentive for the work came with the founding of the Internationl Astronomical Union (IAU). It was set up in 1919 for the purpose of making suggestions toward greater uniformity in nomenclature and to promote cooperation among astronomers.

Within the IAU a commission was set up to establish a classification scheme. G. Bigourdan was selected to head it. Hubble and others presented classification schemes, but when the summary of the meeting was written up and distributed by Bigourdan, only his own and other European schemes were mentioned. Hubble's contribution was ignored. This outraged Hubble, but it did not stop him from continuing to work on the problem. He wrote the president of the IAU, V. M. Slipher of Lowell Observatory in Arizona, telling him of a new scheme. In the letter he emphasized that any such scheme should be simple and based primarily on observational characteristics, but in the back of his mind was a theoretical evolutionary sequence that had been devised a few years earlier by the British astrophysicist James Jeans. (His continual referral to this unproven theory, in fact, eventually caused him considerable grief.) Hubble put what are now known as elliptical galaxies into a classification he

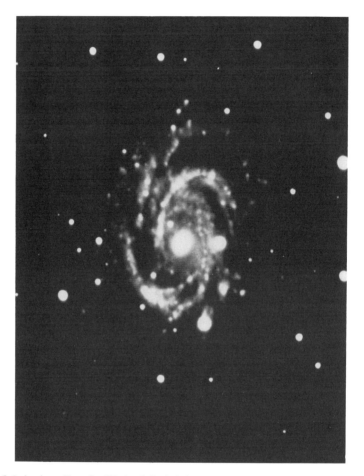

Spiral galaxy, Type Sc. (National Optical Astronomy Observatories, Tucson, Arizona)

referred to as amorphous galaxies. He divided them into classes A0, A1, A2, and A3, according to their shape. In the same way he put spirals into another classification designated by the letter S. He divided them into four groups.

Later in 1923 Hubble wrote Slipher again with a new

Barred spiral. (Hale Observatories)

scheme. It was quite different from the earlier one, and quite close to the one we use today. He now had three main classes: elliptical (previously amorphous), spirals, and irregulars. The ellipticals were designated by the letter E and a number that gave a measure of their shape as seen in the sky. E0 was round,

Elliptical galaxy. (National Optical Astronomy Observatories, Tucson, Arizona)

and E7 (the upper end) was quite elongated. The spirals were divided into two classes, according to whether they had a bar through their center (SB), or had regular spiral arms (S). He further divided the spirals into classes a, b, and c, according to how tightly wound they were.

But again Jeans' theory was on his mind. He said he believed there was some justification for considering elliptical galaxies to be an early stage of spirals. Furthermore, he labeled the spirals "early, middle, and late," indicating a stage in their life. Although a number of astronomers encouraged Hubble to publish his scheme, he decided to wait until the members of the IAU commission on nebulae had considered it. Unfortunately, the next meeting of the IAU wasn't until 1925, so he had a considerable wait. And the reception to his scheme, when it finally came, was generally negative. The major problem, the members felt, was his attempt to relate it to Jeans' theory. They thought that the scheme should be independent of any theory that was still in doubt, and were reluctant to sanction it.

Hubble was disappointed, and when he published the scheme he did not refer to Jeans' theory. The delay also caused other problems. The Swedish astronomer Lundmark had been working on a scheme similar to Hubble's. When Hubble saw it he was annoyed and wrote Slipher complaining that it was plagiarized. He pointed out that Lundmark had been at the 1925 meeting in which he (Hubble) had presented his ideas. Nevertheless, there is evidence suggesting that Lundmark had been working on a scheme for several years.

Despite the problems, the controversy and the final nonacceptance by the IAU, Hubble's classification scheme has been accepted by astronomers. It has been modified slightly over the years, and some additions have been made; nevertheless, it is basically the same scheme that Hubble presented to the IAU in 1925.

PROBING DEEPER: THE BIG EYE

Hubble continued working on various aspects of galaxies. In 1929 he published a paper showing that galaxies are all moving away from one another. This was followed by more conclusive proof in 1931. The universe of galaxies was indeed expanding. He summarized his findings in his book *Realm of the Nebulae* in 1936.

George Ellery Hale.

But if further progress was to be made in understanding the structure of galaxies, and their role in the universe, a larger telescope was needed. George Ellery Hale, the director of Mount Wilson Observatory, had already begun talking about the need for a larger telescope by the early 1930s. He pointed out the advantages; a 200-inch telescope, he said, would have four times the light-gathering power of the 100-inch Hooker telescope. With a little persuasion, the Rockefeller Institute finally agreed to fund it.

Overjoyed, Hale began looking for the ideal spot to mount it. In the spring of 1934 he drove to the summit of Mount Palomar, a peak near Pasadena. After an extensive search he settled on a meadow at one end of it. It was the perfect spot: dark skies free from city lights, and the stability of the atmosphere (called the "seeing") was excellent. After some initial planning the well-known artist and amateur astronomer Russell Porter was hired to draw up plans for both the dome and the telescope. The mirror blank was cast at Corning, in New York. It was a tremendous undertaking; nothing this large had ever been cast before. The first casting was a flop, but the second was successful. The huge glass disk was slowly cooled over a period of 10 months, then shipped to Pasadena.

It arrived in Pasadena in April of 1936, but George Ellery Hale was too ill by then to see the realization of his dream. He spent most of the rest of his life in a sanitorium. When he was well enough he would go to his solar laboratory where he would watch flares and other disturbances on the sun. The sun was his lifelong passion. Two years after the arrival of the mirror he died.

At the time of Hale's death workmen had barely begun figuring the surface of the mirror. The grinding and polishing were done at the Caltech workshop. The grinding continued for 8 years, interrupted only by World War II. Finally, in November 1947, the mirror was finished and it was taken up the steep road to the summit of Palomar. A thin layer of aluminum was deposited on its surface, and just before Christmas of 1947 it was placed in the mirror chamber. One by one, astronomers began to look into the eyepiece. Everyone agreed that the star images were unmatched, but further refinements were still needed before it could be used regularly.

The telescope was officially dedicated on June 3, 1948. It was named the Hale telescope. Routine observations, however, did not begin until November 1949.

Hubble was one of many astronomers looking forward to using the new telescope. He would not only be able to see much farther into the universe, but he would be able to make a much more extensive study of galaxies. He was sure that tremendous breakthroughs were just around the corner. In the summer of 1949, just months before the telescope went into full-time operation, he left for his annual fly-fishing vacation in Colorado. He

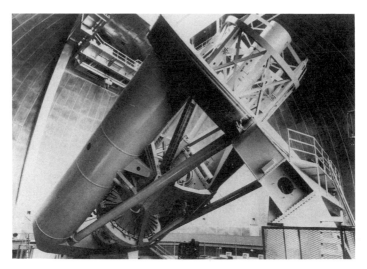

The 200-inch Palomar reflector. (Hale Observatories)

and his wife had traveled to the same ranch each summer for 20 years.

He felt tired as he boarded the train, but he was sure it was just from overwork. He would feel much better once he got on the streams of Colorado. And indeed he did. But then one night he felt sharp pains in his chest. There was no doctor nearby so he had to make an arduous trip to Grand Junction the next morning. Tests were quickly made and, sure enough, he had suffered a massive heart attack. That night he was in considerable pain and his wife worried that he would not make it through the night. But he did, and within a few days he had recovered sufficiently for the trip back to Pasadena.

He was bedridden for several months. Others had now begun to use the 200-inch telescope, and it was a frustrating time for him. He had waited so long. But he continued to recover month after month, and finally in 1953 he was able to use it. By September he felt he was back in his old form and able to work the entire night through.

On a trip to the telescope in late September he took his wife along. When he finished work in the early hours of the morning he said to her, "Come on up and look at the telescope." As they stood there he explained several of its features to her. They then turned and left. It was the last time he ever saw the telescope.

A few days later his wife drove him to his office in Pasadena. After working for a while he decided to walk home. His wife, coming home from shopping, spotted him and picked him up in the car. As she went to pull into their driveway she noticed that he had suddenly leaned forward with a strange look on his face. She quickly stopped the car. "Are you all right?" she asked. "Don't stop, drive on in," he replied. She knew immediately that something was wrong. As soon as she stopped the engine she hurried around to his door to let him out. But by the time she got there he was dead.

Hubble made very little use of the 200-inch telescope, but since his death it has been used to make many important discoveries. As a last tribute to him, his student Alan Sandage gathered together most of his photographs and published them in 1961 under the title *Hubble's Atlas*.

SPIRALS AND ELLIPTICALS: A CLOSER LOOK

Hubble had opened a new door to the universe. Not only do astronomers now know that the fuzzy white patches in the sky are galaxies similar to our own, but they now also have a good idea of the overall structure of the universe. Still, many problems remain unsolved. How did the galaxies come into existence? Why are some spirals, others ellipticals? Why are some extremely energetic, others docile? Astronomers are still struggling with these questions today.

Let's take a closer look at the spirals. What do we know about them? First of all, we know that appearance can be deceptive. When we look at the photograph of a spiral we see two or three prominent spiral arms wound around one another. It appears as if most of the stars are located within these arms, with

Spiral galaxy in the constellation Ursa Major. (Palomar Observatory, Copyright, California Institute of Technology)

few stars between them. But this isn't so. There are many large blue stars and a considerable amount of gas in the arms; this is, in fact, what makes the arms stand out. But in the dark spaces between the arms there are large numbers of small, faint stars—

stars that do not show up in photographs. On the average, in fact, the distribution of stars is fairly uniform around the galaxy.

There are differences in certain regions of the galaxy, however. The central bulge of a spiral is, in many ways, similar to an elliptical galaxy. Take away the outer regions—the arms—and what you have left is basically an elliptical. Ellipticals are composed mostly of tiny stars called red dwarfs, and they have no gas or dust. This is exactly what we see in the core of a spiral.

We saw earlier that there is considerable variation in the form of spirals: Some are loosely wound, others tightly wound. Furthermore, some have a strange bar through their center. Let's look at this spiral shape in more detail. What causes it? One possibility is that spirals rotate, with the inner stars orbiting faster than the outer ones. We might expect this, since we know that the inner planets in our solar system rotate faster than the outer ones. If this were the case we would expect spirals to wind up tighter and tighter as they rotated. This would be like taking a pail of white paint and drawing a line across it with black paint, then stirring near the center. A spiral would, of course, form immediately, and as you stirred it would get tighter and tighter. Does the same thing happen in the case of spirals? If it does, it means that loosely wound spirals (Sc) are younger than moderately tightly wound spirals (Sb), and they, in turn, are younger than tightly wound ones (Sa). Why? Because of the finite speed of light, when we look out into the universe we are actually looking back in time. We are, for example, seeing a galaxy that is 10 million light years away at a much earlier period of its life than when we are seeing one that is only 2 million light-years away. (This assumes, of course, that all galaxies were created at the same time, and we believe that this is the case.)

If the differential rotation idea described above is correct the most distant galaxies would all be Sc. Closer ones would be Sb and the closest Sa. But this isn't the case. Galaxy types are roughly uniformly distributed.

But if differential rotation isn't responsible for the spiral shape, what is? The theory that is now generally accepted is

called the density wave theory. It was introduced by Bertil
Lindblad of Sweden and developed mathematically by the
American astronomers C. C. Lin and Frank Shu in the 1960s. In
this theory a wave is assumed to move through the outer regions of a galaxy. The wave has little effect on the stars, but
pulls gas and dust into certain regions and out of others. More
gas is attracted gravitationally to the dense regions caused by
the wave. Clumps build up and, in turn, are attracted to one
another, until finally stars begin to form along the "wake" left by
the wave. Some are bright, and excite the gas around them.
From a distance a string of such stars and gas looks like a spiral
arm.

But the central bulge and spiral arms are not all there is to a
spiral. Evidence accumulated over many years indicates that
spirals are surrounded by gigantic clouds of what astronomers
call dark matter. This is matter that cannot be seen or detected
directly. It is detected only through its gravitational pull. But
without this halo, galaxies would be unstable and would soon
break up. We're still not sure what the dark matter is composed
of, but many astronomers are currently working on the problem.

Finally, surrounding the galaxy, like bees around a hive, are
globular clusters. They are independent systems of stars usually
containing a few hundred thousand stars; they vary in diameter
from 15 to 300 light-years. Globular clusters do not follow the
rotation of the galaxy, but have their own orbits. Much of the
time they are outside the disk of the galaxy but twice in each
orbit they pass through it.

In conclusion, galaxies are obviously complex and intriguing objects. And although much has been learned about them in
the past few decades, there is still much to be learned.

CHAPTER 3

The Discovery of Radio Sources

In the last chapter we saw that the white fuzzy objects in the sky were eventually shown to be island universes of stars—galaxies. Some of these galaxies are spirals, others ellipticals, and a few have little or no form. But galaxies differ in more than just form. Some are much more active than others, with cores that eject energetic radiation and particles. To us they appear as if they are exploding, for we frequently see long plumes of gas and stars, and in a few cases we even see jets emanating from them. Most galaxies, however, show little evidence of activity. Our Milky Way galaxy, for example, would show none if viewed from a distance. A close look at its core, however, reveals that it too is energetic. How do we know? Although optical telescopes don't show it, other types of telescopes do.

One of these other types of telescopes is the radio telescope. Radio waves are, of course, invisible to our eyes, nevertheless you are no doubt familiar with them. As their name implies they are the waves that allow us to listen to radio (and watch television). Their existence was predicted in 1864 by Clerk Maxwell, and they were discovered in 1888 by Heinrich Hertz.

How do radio waves differ from light waves? Both are part of what is called the electromagnetic spectrum. Ultraviolet, infrared, and microwaves along with X rays and gamma rays are also part of this spectrum. Each of these waves is characterized by a wavelength (distance between two adjacent humps on the wave). Radio waves, for example, have wavelengths that are

41

typically a few feet long (but they can be as short as an inch). Light waves, on the other hand, have a wavelength roughly a million times shorter.

JANSKY

Only a few years after Hertz discovered radio waves, the Italian engineer Marchese Marconi sent the first radio message across the Atlantic (1901). Within a few years radio-communication techniques had improved considerably. Most of the early work was done at long wavelengths. Then it was discovered that short waves worked better at long distances. And in 1927 AT&T began offering shortwave radio telephone service across the Atlantic: three minutes for $75. Radio waves were, by then, well understood; scientists, had, in fact, even checked to see if the sun was a radio source. But they found it to be radio-silent.

It probably never occurred to anyone at the time that radio waves might be coming from beyond the sun. But in 1929 Karl Jansky of Bell Telephone Labs showed that this was, indeed, the case. Like many of the later discoveries in radio astronomy it was an accidental one. Jansky had no idea where the waves were coming from when he first detected them; they were just a strange hiss in his antenna. But he wasn't one to let a challenge go unheeded. What was causing the hiss? Where was it coming from? He was determined to find out.

Born in 1905 in Oklahoma Territory, Jansky was the third of six children. His father was an electrical engineer who taught at various colleges in the Midwest. Although encouraged by his father to go into engineering, Jansky chose physics instead. He did, however, get some experience in electronics; his senior thesis was on vacuum tubes. He graduated in 1927 but stayed on for a year while he completed courses for a master's degree (he didn't complete his thesis until several years later).

In 1928 he applied for a job at Bell Labs in New York City. During his physical exam doctors discovered that he had a se-

Karl Jansky and his antenna. [National Radio Astronomy Observatory (NRAO)]

rious kidney disease. Jansky, of course, already knew of the disease, as it had disqualified him from ROTC two years earlier. Despite it he had been active in sports, one of his favorite being ice hockey. He played on the University of Wisconsin hockey team. Although Jansky had never let the disease slow him down, Bell Lab officials were concerned that it might limit him, and they were reluctant to offer him a job. Fortunately, his brother, a professor of engineering at a nearby college, had connections with Bell Labs, and he assured them that Karl's potential far outweighed any shortcomings that might result from the disease. He was therefore hired.

After some initial training Jansky was sent to Cliffwood, N.J. to work under Harald Friis. The first shortwave telephone calls across the Atlantic had just been made but static was interfering with the signal. Jansky was assigned the task of finding out what was causing it. It was a challenging problem, mostly because his background in radio and electronics at this stage was still weak. "Before I came here, the language they spoke was almost foreign to me," he wrote to his father. "But I'm beginning to get used to it now. At Madison I never heard of such things as attenuators, T.U.'s, gain controls, double detection, etc., but that's what I get for not taking engineering."

Jansky's first job was to build a large, movable antenna which would allow him to detect the static. Made of brass pipe and wood, it was 115 feet long when completed. Model T wheels were used to turn it on its circular track. He had barely gotten it assembled and working when Bell Labs decided to move their Cliffwood operations a few miles away to Holmdel, N.J. This caused a delay, but it meant that Jansky would now have much more room for his antenna. Furthermore, Holmdel was a more isolated area and there was less man-made radio interference.

Once his antenna was reassembled he began taking data. Convinced that there would be little static during the winter he left the project for several months over the winter of 1930, then began again early the following spring. Soon he had enough data to see an emerging pattern. There appeared to be three distinct types of static:

1. Relatively intense static which was traced to local thunderstorms.
2. A residue of less intense static due to distant thunderstorms.
3. A strange hiss, weaker than the others, that was periodic through the day.

It was the third type of static that aroused the most interest in him. He had no idea what it was, but he knew it was quite different from the other types, so he decided to investigate it. At first he thought it might be coming from the sun; he even began referring to it as "sun static." But gradually it moved away from the direction of the sun. Still, it occurred roughly once a day; in other words, it was periodic with a period of approximately 24 hours. He wrote a letter to his father describing it, ending the letter with, "Sounds interesting, doesn't it?"

As he continued studying the static he noticed that the direction it was coming from gradually shifted across the sky; it seemed to be associated with the seasons. Whatever it was, Jansky realized, it was important. And apparently his supervisor Friis agreed, for he suggested that Jansky write a paper on it for publication. Friis also suggested that he present the paper at the annual meeting of the International Scientific Radio Union (URSI) in Washington, D.C. "It was not my wish to present the paper there," wrote Jansky. "I wanted to present it at the IRE (Institute of Radio Engineers) convention in Chicago in June, but Friis said no. The URSI meetings in Washington are attended by a mere handful of old college professors and a few . . . engineers." The reception to his talk, as he expected, was lukewarm. "Not a word was said about my [paper] except for a few congratulations that I received afterward." Fortunately, enough attention was generated for him to get an invitation to present it at the IRE convention in Chicago.

In December 1932 the astronomical significance of the signals finally began to come to light. A fellow worker suggested that Jansky plot up the data he had, which were about a year's worth. They showed that the source of the waves appeared to be a fixed plane.

Jansky then began talking to Melvin Skellett, an employee of Bell Labs who was also a graduate student in astronomy at Princeton University. Skellett told him about celestial coordinates, astronomical time, and so on. This was the key, Jansky realized. He immediately got some books on astronomy and began studying them. And soon he was convinced that the radio waves were coming from the direction of the Milky Way. They were, in fact, coming from the center of our galaxy.

In his second paper, which was published in 1933, he announced his discovery. But Friis remained skeptical and cautioned him to be careful in writing up the report.

Then something quite unexpected happened. Soon after the publication of the second paper Bell issued a press release. A few days later the front page of *The New York Times* contained the heading "New Radio Waves Traced to the Center of the

Milky Way." Jansky was suddenly a celebrity. The "star noise" was even broadcast over a nationwide radio station.

But what was causing the noise? If it was coming from the Milky Way, it was reasonable to assume that it was generated by stars. After all, the Milky Way is made up of stars. Yet when Jansky turned his antenna toward the sun it was radio-silent. Stars, it seemed, couldn't be causing it. But there was a considerable amount of interstellar gas between the stars. Could it be causing the noise? Strangely, no one looked into this possibility.

It would, in fact, seem that, with all the publicity, astronomers would begin building their own antennas to study the new phenomenon. But this didn't happen. In fact, shortly thereafter, in January 1934, Jansky was taken off the project and assigned to a new one. But his interest in the star noise didn't die, and he kept hoping he would eventually get back to it. Through 1934, however, he did no further work on it. But early in 1935 a senior engineer suggested that he write another paper. And he did—but it was based on work he had completed two years earlier. This, in fact, was his last paper. He persisted in his attempts to get someone interested in the discovery; he even submitted a proposal for a 100-foot steerable radio antenna, but it got lost and was never built.

Jansky tried to get astronomers interested by writing an article for *Popular Astronomy*. And indeed, Harlow Shapley of Harvard wrote him and even considered the possibililily of a project at Harvard. But nothing came of it. In 1935 two scientists at Caltech verified the discovery, and Fritz Zwicky encouraged Caltech to build a radio telescope. But again, nothing was done. It was somehow too far removed from what astronomers usually studied, and they took little interest in it.

With the United States' entry into World War II, Jansky was assigned to war-related projects and never got back to his "star noise." By the end of the war his health was deteriorating rapidly. He spent some time in Duke University hospital and was placed on a strict diet. But nothing seemed to help. In 1950 at the age of 44 he died of kidney and heart failure. The new era of radio astronomy was just beginning.

REBER

Although there was little interest in the discovery from professional astronomers, interest did not die. Grote Reber, an electrical engineer at Wheaton, Illinois, read about the discovery and soon became fascinated with it. He decided to build a radio antenna to study the phenomenon.

Born and raised in Wheaton, a suburb of Chicago, Reber graduated from the same high school Hubble did. His mother, in fact, was Hubble's teacher. While on a trip to Hawaii in 1952 Reber stopped off at Pasadena and visited Hubble.

"Do you remember your 7th and 8th grade teacher?" he asked Hubble.

Hubble nodded. "Yes, it was Miss Grote."

"She's my mother," said Reber. He then went on to tell Hubble that she had followed his career for years. She had, in fact, given him (Reber) one of his books when he was young.

Reber's interest in radio had already blossomed by the time he was 15. He had built a transceiver and talked to people in almost every country of the world via shortwave radio. He got his B.S. in electrical engineering from what is now the Illinois Institute of Technology in 1933. Upon graduation he began working for an electronics firm in Chicago.

"It was obvious that Jansky had made a fundamental and important discovery," Reber wrote shortly after seeing Jansky's paper. He was sure there was much more to be learned, and the only way to learn it was to build a radio antenna capable of detecting the waves. He therefore wrote to several astronomers suggesting that they build the antenna while he supplied the electronics expertise. But no one was interested. Undeterred, he decided to get it built, and pay for it out of his own pocket. Taking the design to several engineering firms, he found their

Grote Reber and his original antenna. [National Radio Astronomy Observatory (NRAO)]

quotes to be far over his budget. The only alternative was to buy the materials and build it himself.

The antenna dish was to be parabolic in shape, so that the waves would all come to a focus at a single point. He chose the focal length to be 20 feet, and decided on a dish diameter of 31 feet. In June 1937, he began building the device, and in September of the same year he finished it. The total cost was $1300—a considerable sum for that time.

People began to wonder what the monstrosity in his backyard was. No one had ever seen anything like it. Adding to the intrigue was the fact that it snapped, popped, and cracked every morning and evening due to thermal expansion. And when it rained, water poured out of the hole in its center. People began to think it was some sort of meteorological device. Planes would occasionally swoop down over it to have a better look. One day, in fact, a low-flying plane developed engine trouble over it and had to land in a nearby field. Neighbors began to wonder: Was it some sort of secret war weapon?

Reber started his search for the radio noise by tuning his antenna to relatively short waves (approximately 3 inches). He pointed it at the sun, moon, planets, and bright stars. Nothing. There was no noise; each of the objects appeared to be radio-silent. He therefore changed to a longer wavelength (approximately 13 inches) and tried again. This time he took both day and night observations; he checked everything, but again everything was radio-silent. The only thing left to do was try an even longer wavelength; he therefore tried a wavelength of approximately six feet.

With the new, longer wavelength he finally detected "static" from the Milky Way. But it soon became obvious that he would have considerable difficulty studying the static. During the day hundreds of automobiles passed his house, and each of them generated radio static. It was virtually impossible to separate the weak signal from the Milky Way from all this automobile noise. Only at night was his antenna quiet enough for serious study of the star noise.

But Reber worked all day 30 miles away in Chicago. If he was to also work with his antenna at night he would have to carefully plan his time. He therefore set up a rigid schedule. Upon returning home from work at around 6 P.M. he would eat, then sleep until midnight. At midnight he would rise and record observations until dawn. Then he would go to work in Chicago.

Day after day he continued this backbreaking schedule, un-

Grote Reber in later life.

til finally he had enough data to publish. His first paper appeared in the *Proceedings of the IRE* in February 1940. In it he described the noise and suggested a possible source. He believed it was caused by interactions between electrons and ions (heavier charged particles) in the gas clouds of the Milky Way.

But Reber wanted to attract the attention of astronomers. After all, it was an astronomical phenomenon. He therefore sent a paper to the *Astrophysical Journal*. As in the case of all such papers it was sent to two referees for their comments. Both sent it back saying that it was unacceptable for publication. But the editor, Otto Struve, wasn't convinced; he decided to look into it further. With another astronomer, Gerard Kuiper, he therefore visited Reber and his radio telescope. Interestingly, the visit took place on a Monday, and Reber's mother was using part of the telescope as one end of her clothesline, so it couldn't be moved.

A demonstration was, however, arranged, and after seeing it Struve decided that the paper deserved to be published. He arranged for two theoreticians, L. G. Henyey and P. C. Keenan of Yerkes Observatory, to develop a theory for the source of the waves, and their paper was published along with Reber's. Their predictions, in fact, agreed quite well with Reber's observations.

Reber was now ready for bigger things. He decided to make a complete map of the Milky Way. "The success whetted my appetite," he said. "If a little is good, more is better." He bought an automatic recorder and began the survey. After collecting about 200 chart recordings of radio noise, he plotted them up. The intensity peak, as expected, was in the direction of the center of the Milky Way, but strangely, there were two secondary peaks—one in the constellation Cygnus, and one in Cassiopeia. This, incidentally, was the first published radio map of the Milky Way.

Oddly enough, during all the time he was working on the Milky Way, Reber never tried to detect radio waves from the sun. He had tried earlier and found nothing. Finally, however, in 1943 he turned his antenna toward the sun and was surprised when the needle went off scale. The sun was a source of radio waves, after all—a strong one. But by now the discovery had already been made in England.

WORLD WAR II AND RADAR

During World War II England relied heavily on radar to provide advance warning of air raids. Army personnel were therefore shocked in 1941 when radar jamming from the coast of France was so effective that two battleships managed to get through the English Channel virtually undetected. A short time later intense jamming of radar occurred again. Everyone braced for an air raid, but nothing happened. Wondering what was going on, the

British army assigned James Stanley Hey, a physicist associated with their scientific research group, to look into the problem. He began by talking to several of the radar operators. They all agreed that the jamming seemed to be coming from the direction of the sun. Hey considered the coincidence. Was it possible that the jamming wasn't due to the enemy? He knew that the sun went through an 11-year sunspot cycle, and during the sunspot maximum there was a considerable amount of activity. Was the sun causing the jamming? He phoned the Royal Observatory. They said the sun was still two years from maximum, and little activity was expected, but surprisingly a large sunspot had appeared a few days earlier. This convinced him; he was sure the sunspot had caused the jamming. Sunspots have strong magnetic fields, and radio waves are generated when electrons pass through them.

Hey wrote up a paper on his findings, but it was not published until the end of the war.

POST-WORLD WAR II: THE EARLY YEARS

When the war was over Hey decided to study the radio emissions from the sun in more detail. There were, in fact, many experts in radar who wanted to use their talents in the field. Hey therefore organized a government research group to study the radio waves from the sun. A group was also set up under Bernard Lovell at Manchester University. Lovell's main interest was cosmic rays (charged particles that strike our atmosphere creating other particles and radiation). He was convinced that with increased sensitivity he would be able to detect cosmic ray "echoes," using a radio antenna.

Lovell obtained his doctorate at the University of Bristol in 1936. Upon graduation he began teaching and doing research on the ionosphere and cosmic rays at University of Manchester. Like many other scientists he worked on radar during the war. During his research he discovered that meteors, or "falling stars," showed up well on radar. After the war he therefore decided to build an antenna to study meteors, and (he hoped) cosmic rays.

His first antenna was a giant spider web of steel, 218 feet across. It was mounted on the ground and, as a result, could only detect sources that passed directly overhead. Nevertheless, Lovell and his group did a considerable amount of important work on meteors with it (but he never did detect cosmic rays). Eventually, though, Lovell began thinking about a giant movable dish—an antenna that could be pointed in any direction in the sky. He decided to build one.

Meanwhile, at Cambridge University, another group was making considerable progress under Martin Ryle. Ryle graduated in physics from Oxford in 1939, and, like Lovell, worked with radar during World War II. One of the first projects of his group was a extensive survey of radio sources. They eventually developed techniques for getting accurate estimates of the positions of sources.

A similar group was formed in Australia under Joseph Pawsley. As we will see later, some of the most impressive early results came from Australia.

JODRELL BANK

After some success in detecting meteors with his 218-foot fixed antenna, Lovell approached his superior, Patrick Blackett, with a proposal to build a 300-foot fully steerable radio telescope. Lovell's timing was perfect. The war effort had convinced Blackett that the future of science was going to be centered around large intruments. Without hesitation he agreed to back Lovell. Lovell therefore started planning. His first hurdle was finding an engineering firm that would design and build the telescope. He approached several, but they all turned him down. Then, in September 1949, he met H. C. Husband, an engineer who was visiting the 218-foot fixed antenna. He told him his trouble. "I

Bernard Lovell.

want a telescope of this size," he said, pointing to the antenna
on the ground, "that we can steer to every point of the sky. I've
been trying for over a year to persuade someone to do it but I'm
told it's impossible."

Husband looked over the antenna for a few moments, then
said, "It should be easy—about the same . . . as throwing a
swing bridge over the Thames." Lovell could hardly believe
what he had heard. He had finally found someone who believed
the telescope was possible.

Husband drew up some preliminary plans and took them to
Lovell. As Lovell examined them he realized that he was dealing

with an engineer of considerable competence. With Blackett's backing he got funding from the Department of Scientific and Industrial Research (DSIR) and the Nuffield Foundation. He was sure that the telescope would be completed within a short time—perhaps as early as 1954. But then came the first setback: Several things had not been included in the original estimate and he would need more money. Then there were problems getting the land he wanted. But Lovell was determined to build the telescope, and nothing was going to stop him. Finally it was time to drive the first piles for the foundation. It was expected to take two months. It took nine. Then the price of steel sky-rocketed, and more funds were needed. Lovell wasn't sure where he would get them. Fortunately, DSIR finally agreed to cover the overruns.

Troubles continued to compound. Project after project was delayed until by 1954 (Lovell's original estimate for when it would be completed) only the foundation for the telescope had been constructed. Then a major change in design was needed. The original dish had been designed to detect only long wavelength radio waves, but in 1951 short waves (21 cm) had been detected coming from the Milky Way. Lovell wanted the telescope to be state-of-the-art. Husband therefore had to make several changes in the design.

In 1955 the two towers that were to hold the dish finally began to rise. But the cost of steel took another jump and Lovell's debt rose to almost half a million dollars. He had no idea where the money would come from. "There were many periods of despair during the years of construction," said Lovell, "but spring of 1955 was one of the blackest." Lovell had to appear before the Public Accounts Committee and try to explain the huge debt. For a while he feared he might even be imprisoned.

Then something remarkable happened. In October 1957, Sputnik was launched by the Russians. The world was galvanized. Signals from the satellite were relatively easy to detect, but no one had detected the launching rocket. Newpaper reporters and broadcast personnel appeared at Jodrell Bank; soon there were hundreds of them. Lovell had never seen anything

The Jodrell Bank radio telescope. Now called the Lovell telescope. (Nuffield Radio Astronomy Laboratories, Jodrell Bank, Cheshire)

like it. He wasn't sure what to do. The telescope wasn't ready, and it couldn't be operated from the main console. But an offer of a cable soon came from an engineering firm, and it was quickly hooked up.

With hundred of reporters around, the telescope was slowly pointed in the direction of the satellite. Everything worked perfectly as radio echoes from the rocket were searched for and found.

It was a tremendous success for Lovell. The feat was announced in the House of Commons by the Prime Minister and Lovell's money worries were soon over.

Since that time the Jodrell Bank dish has been used to study many astronomical phenomena, including meteors, planets, the sun, and galaxies. It is no longer the largest fully steerable radio telescope in the world, but it has played an important part in the history of radio astronomy. For his contribution in building it Lovell was knighted in 1961.

WORK AT CAMBRIDGE

While the Jodrell Bank dish was being built, considerable progress was being made at Cambridge University. A major breakthrough was the use of interferometry in radio telescopes. One of the main problems with radio telescopes was their poor resolution (ability to distinguish clearly two closely spaced objects). Resolution is related to wavelength and, because the wavelength of radio waves is so long, it would take a dish 150 miles across to equal the resolution of the eye at night. With single dishes the view is so blurred it is virtually impossible to pinpoint radio sources. This meant that it was impossible to associate a given radio source with an optically identifiable object in the sky.

It was discovered at Cambridge (and about the same time in Australia), however, that if two radio telescopes were linked, and their signals brought together in the right way, the combination could be made equal to a telescope with a diameter equal to the separation.

Using a linkage of this type, called an interferometer, Cambridge astronomers began a survey of radio sources in the sky in the early 1950s. Techniques were still crude, but they were able to pinpoint sources reasonably well. Their first survey, called 1C, produced a catalog containing about 50 entries. But it was short-lived. As more sophisticated interferometer techniques were developed, a second survey was initiated. It was much more ambitious, and when it was published in 1955 it contained over 2000 sources. But similar work was going on in Australia, and they could not confirm many of the sources. Something was obviously wrong! Detailed checks were made and they soon showed that many of the sources were spurious. The Cambridge team therefore had to start again, and in 1959 they pub-

Martin Ryle.

lished the 3C catalog. It contained 471 sources and is still used extensively today. (Sources from the survey are referred to by 3C plus a number, e.g., 3C 48.)

Martin Ryle and his group at Cambridge continued to perfect interferometry techniques. In 1962 they built the Mile Long Radio Telescope which consisted of three 60-foot dishes linked along a line. The middle of the three could be positioned anywhere between the other two. But if they were to approach optical telescopes in resolution the dishes had to be separated much more. And soon they were. The first interferometers were linked by cable; later, radio frequency links were used and the separation increased. Finally, Very Long Base Interferometry (VLBI) was developed. It became possible because of the development of atomic clocks, high-speed computers, and accurate magnetic recording tape. In this case tapes are made at two stations, frequently several thousand miles apart. Atomic clocks are used to record the incoming waves accurately. The tapes are then brought together and superimposed to produce a high-resolution image.

AUSTRALIA

While England was busy surveying the northern skies, Australia was surveying the southern skies. Like Cambridge, they were also using interferometry. A particularly ingenious interferometer, in fact, was developed and used by John Bolton. He perched a radio dish on a cliff overlooking the Tasman Sea. Reflected radio waves from the water could then be picked up along with signals directly from the source. The two sets of waves could then be superimposed to produce a sharper image. Bolton used this technique to identify several sources optically.

Trained at Cambridge, Bolton served with the British Navy in the Pacific. When the war was over he decided to remain in Australia. With his interferometer he was able to identify three sources that he referred to as Taurus A, Virgo A, and Centaurus A. Taurus A turned out to be the Crab Nebula, Virgo A, the exploding galaxy M87, and Centaurus A was identified with galaxy NGC 5128. The Crab Nebula is a well-known supernova (exploding star) remnant. Over the years it has become one of the most studied objects in astronomy. M87, an elliptical galaxy, is particularly interesting because it has a remarkable jet emanating from its nucleus. Centaurus A is a strange-looking galaxy with a dark band of dust cutting through it.

One of the most interesting sources in the sky at the time was Cygnus A. Bolton narrowed in on it but was unable to identify it with an optical counterpart.

Australia soon became famous for its huge cross antennas. They are two long lines of small antennas placed perpendicular to one another. The technique was discovered by B.Y. Mills in 1952 and in 1953 the 1500-foot Mills Cross was built. Australia also soon had its counterpart to the Jodrell Bank Mark I. A fully steerable 210-foot radio dish was completed in Parkes in 1961.

THE UNITED STATES

After a promising start with the early work of Jansky and Reber, radio astronomy began to lag in the United States. The only important work during this time was performed at the Naval Research Labs in Washington, D.C. In 1950 radio astronomers completed a 50-foot dish, and a number of important discoveries were made with it, including the detection of radio emission from gaseous nebulae and several planets.

By the late 1950s astronomers in the U.S. began to realize that in the field of radio astronomy they were lagging seriously behind England, Australia, and even Holland. Shortly thereafter the National Radio Astronomy Observatory was organized, and an 85-foot dish was built at Green Bank, Virginia. In 1963 Cornell University completed a 1000-foot fixed antenna in Puerto Rico. It was placed in a natural hollow in the countryside.

Plans were then made to construct a huge assembly of telescopes, later called the Very Large Array (VLA). Three 85-foot dishes were built at Green Bank to test the feasibility of such a huge array. After three years of experimentation astronomers were ready to go. A site was selected near Socorro, New Mexico. The VLA was to consist of 27 dishes, each 82 feet in diameter. They would be put in a Y configuration with each leg 13 miles in

A section of the Very Large Array (VLA) at Greenbank, Virginia. [National Radio Astronomy Observatory (NRAO)]

length. The signals from the dishes, brought together by computer, simulated a dish 25 miles in diameter. The project was completed in 1980.

Large radio telescopes have now been constructed in many centers of the world, including Holland, Russia, Canada, Germany, and Japan. The largest fully steerable dish in the world at the present time is in West Germany. The largest fixed antenna is the 1000-foot one at Arecibo, Puerto Rico. Several others, including one to be called the Very Long Baseline Array (VLBA), are in the planning stages. Observations from these instruments will no doubt prove to be exciting challenges that will occupy astronomers for many years to come.

CHAPTER 4

Exploding and Peculiar Galaxies

By 1950 it had become evident that many of the radio sources in the sky were compact (of small angular diameter). But astronomers still had no idea what most of them were. Bolton had identified Taurus A with the Crab Nebula, which was known to be a supernova remnant. Did this mean that most of the other radio sources were supernova remnants? It seemed unlikely. In fact, the other two sources that Bolton had identified were both galaxies. Furthermore, one of the strongest sources in the sky—Cygnus A—had still not been identified optically. Despite the controversy Bolton and his group began referring to all of the objects as "radio stars."

The difficulty of identifying radio sources with optical objects in the sky related to the wavelength of the radio waves. Because it was so long, compared to visible light, the radio sky was extremely fuzzy compared to the optical sky. A way had to be found to get around this problem, and the key seemed to be interferometry. Astronomers therefore put a considerable effort into improving interferometer techniques during the 1950s and early 1960s.

The two major groups studying interferometry in England were Ryle's group at Cambridge, and the one at Jodrell Bank, supervised by Lovell. As one might expect, there was a keen competition between them. Ryle's group, for the most part, followed Bolton in referring to the sources as radio stars. They were convinced that the majority of them would eventually turn out to be local objects.

Jodrell Bank astronomers, on the other hand, noticed early on that the sources were distributed evenly across the sky, as galaxies are. To them, this appeared to indicate that most were beyond our galaxy (extragalactic)—probably radio galaxies. If they were local, they would be in the plane of the Milky Way. Cambridge, however, was not convinced, and tended to ignore the work being done at Jodrell Bank.

Ironically, it was one of the Cambridge radio astronomers, Graham Smith, who initiated a series of events that proved the Cambridge group wrong. Smith joined the Cambridge group in 1946. (When the war ended he still had a year to complete on his physics degree so he finished it first.) Like many others at the time, he soon became intrigued with Cygnus A. What was it? If he was to find out he would have to get a much better "fix" on its position. Bolton had narrowed its position down considerably, but not enough for an optical identification.

In 1948 the Cambridge group completed a new interferometer. Smith decided to use it to take a closer look at Cygnus A. As he narrowed in on it he found, to his surprise, that there was another strong source nearby, in the constellation Cassiopeia (it is now called Cassiopeia A). He also found, to his disappointment, that the new interferometer was not good enough for an optical identification of either object. He and his colleagues therefore designed and built a better interferometer. Like most of the early interferometers it was built of surplus materials and captured German radar equipment. It was a highly successful interferometer in that it gave them the 1C catalog but, again, Smith found that it was still not good enough to give an accurate position for Cygnus A. What they needed was a longer distance between the antennas (a longer baseline). Smith and Ryle therefore decided to build such an interferometer using two captured German Wurzberg radar dishes. They were linked and separated by about 280 meters.

Before trying to zero in on Cygnus A, Smith tried to mea-

Graham Smith.

sure the parallax of several of the sources. If he could do this he would be able to determine their distances. Early astronomers used this technique to determine the distance to several nearby stars. After selecting what they believed was a nearby star, they would record its position against several distant background stars. Six months later when the Earth was on the opposite side of the sun they would record its position again. If it was, indeed, a nearby star, it would appear to move relative to the background stars. You're likely familiar with the technique on a smaller scale. Hold your arm out with one finger up, then blink your eyes back and forth. Notice how your finger appears to move back and forth relative to the background wall. This is parallax. By measuring the associated angles you can determine the distance between your eye and your finger. In the same way astronomers can determine the distance to nearby stars.

Smith hoped to use this technique to determine the distance to Cygnus A and other sources, but he soon realized that it was a hopeless case. The new interferometer did, however, allow him to get an accurate position for both Cygnus A and Cassiopeia A. He was able to narrow them down to a small square about 1 minute of arc on each side (i.e., 1/60 of a degree). Pleased with his success he went to R. O. Redman, the director of the Cambridge Observatory, and suggested that an optical search be initiated. Redman agreed, and assigned David Dewhirst to search through the Cambridge photographic plates that had been taken of the region. He was to follow this up with a telescopic search. But the available plates were inadequate and the weather was so uncooperative that it was impossible to observe anything. Dewhirst did find a tiny wisp at the position of Cassiopeia A, but was unable to observe it directly. Because of the problems, Redman suggested that Smith send the coordinates to Walter Baade and Rudolph Minkowski at Palomar Observatory in California. They had access to the 200-inch reflector, and if anybody could detect an optical counterpart, they could.

In the fall of 1951 Smith therefore wrote them a letter, enclosing the coordinates and asking if they would search for anything that might be giving off a considerable amount of radio emission.

Baade and Minkowski were both heavily involved with other projects when the letter arrived from England. But it was a request that they could hardly ignore. The radio source Taurus A had been identified with the Crab nebula only a few years earlier and Minkowski's specialty was supernovae. Baade had also worked with supernovae. Along with Zwicky, in fact, he had identified the phenomenon and coined the word supernova. They showed that although supernovae were rare in a given galaxy, they were relatively common statistically.

Born in Germany in 1893, Baade had originally planned on going into the ministry. In high school, though, he became fascinated with astronomy. After obtaining his Ph.D. in astronomy from the University of Göttingen in 1919, he taught at the University of Hamburg. In 1931 he came to the United States and began working with the large telescopes at Mount Wilson. A

Walter Baade.

delicate and somewhat nervous person, he had a serious hip ailment; furthermore, one of his legs was considerably shorter than the other, causing him to limp badly.

A few years after he arrived in the United States, Baade got a letter from Minkowski. Nazism was on the rise in Germany, and Minkowski was beginning to feel uncomfortable. He asked Baade if there were any available positions in astronomy in the United States. Baade inquired and was able to get him one at Mount Wilson.

Baade and Minkowski both loved astronomical challenges, and the information sent by Smith filled the bill. Within days they had made several photographs of the region around Cassiopeia A. In October Baade wrote back to Smith, "I would like to let you know that my search for the Cassiopeia source at

Rudolph Minkowski.

the 200-inch has turned up an extensively interesting object close to the measured point. It is an emission [bright] nebula." Like Taurus A, Cassiopeia A turned out to be a supernova remnant. It appeared that the Cambridge group was, perhaps, right after all. Most of the sources might be within our galaxy.

Baade and Minkowski then turned their attention to Cygnus A. They photographed the area and were amazed by what they saw. (Strangely, they didn't write to Smith about their discovery until the following April.) In his reply Baade said, ". . . I photographed the field at the 200-inch last fall after you sent me your accurate points. The result was very puzzling. At the place you gave me there is a rich cluster of galaxies and the radio position coincides closely with one of the brightest members of the cluster. This galaxy . . . is a queer object. In fact, the 200-inch photo suggests strongly that we are dealing with two galaxies which are in actual collision."

A few weeks after Baade wrote the letter Minkowski was able to get a spectrum. It was an exceedingly dim object so it was a difficult task. Finally, though, with a several-hour exposure Minkowski succeeded. Both men were shocked at the result. The redshift of the lines due to the Doppler effect was huge. There was no doubt that the object was a galaxy. It was traveling away from the Earth at a speed of 15,000 km/sec and, according to Hubble's plot of speed versus distance, the object was a billion light years away from us. Amazed as they were with this result, they were equally amazed by the form of the lines. Many of them were emission, or bright, lines. Emission lines only come from hot glowing gas, and such lines had not been seen before in galaxies. This convinced them that it was indeed two galaxies in collision, for only in such a violent event would the gas be heated to temperatures high enough to give emission spectra.

Cygnus A and Cassiopeia A had been identified, but astronomers were now more mystified than ever. One of the two was a supernova remnant, and the other appeared to be two galaxies in collision. Of the known radio sources, two were now supernova remnants and three were galaxies. There was therefore still no consensus as to what most of the objects would turn out to be.

COLLIDING GALAXIES

Most astronomers were excited at first about the possibility of colliding galaxies. But how likely was it for two galaxies to collide? After all, it was well-known that the probability of two stars colliding was extremely low. A comparison of the two situations was made, and it turned out that the probability of a

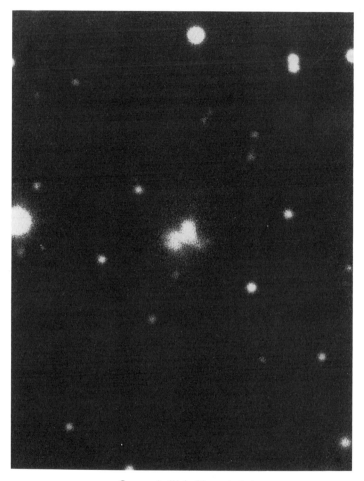

Cygnus A. (Hale Observatories)

collision between two galaxies was much greater than that for stars.

The reason why stars seldom collide is because the distance between them, compared to their diameter (i.e., the ratio of these numbers), is about 30 million. This means that stars occupy a relatively small fraction of the space around them. Furthermore, they are in galaxies, and the motion within galaxies is regular: All stars at a given distance from the center orbit have the same orbital speed. The situation in the case of galaxies, however, is quite different. We are, of course, dealing with a much larger scale, but the important thing is how much space they take up compared to the space between them. A typical galaxy is about 100,000 light-years across (this is the diameter of our galaxy) and a typical separation is two million light-years (the distance to the Andromeda galaxy). The ratio in this case is only 20 (compared to 30 million in the case of stars). Actually, for our local group of galaxies this number is even higher. We took the distance to the Andromeda galaxy as a typical separation, but our nearest neighbors, the Magellanic Clouds, are only 200,000 light-years away. So, any way you look at it the number is small.

This means that the probability of collision, particularly within a cluster, is relatively high. A simple calculation shows, in fact, that we could expect two collisions to be going on at any time in a large cluster such as the Coma cluster.

On the basis of this it seemed reasonable that Cygnus A was two galaxies in collision. And it meant that perhaps many, even most, radio sources might be galaxies in collision. But when the details of the collision were looked at for the case of Cygnus A, astronomers were stumped. Cygnus A was radiating a million times as much radio energy as a normal galaxy such as the Milky Way. It seemed impossible that a collision of two ordinary galaxies could provide this much energy. Stars, on the average, are a long way apart and even if the two galaxies did collide, few if any of the individual stars would collide. The two galaxies would just slide through one another. The only collision that would occur would be the gas clouds of the two galaxies. And there was no way that the collision of two gas clouds could produce a million times as much energy as our galaxy. A head-on collision of billions of stars would have to occur to produce so much energy. And, as far as anyone knew, this was virtually impossible.

Then another problem developed. As work continued, radio astronomers found that the radio waves emitted from Cygnus A and other radio galaxies were polarized. (A wave is polarized when all the photons vibrate in the same direction.) This implied that the radiation was what is called synchrotron radiation—a type of radiation that is caused by high-speed electrons passing through magnetic fields. Colliding galaxies could not produce synchrotron radiation.

THE DISCOVERY OF LOBES

Then another discovery confused the issue even more. With the completion of a new interferometer in 1952 Cambridge radio astronomers Roger Jennison and M. K. Das Gupta decided to take a closer look at Cygnus A. Astronomers had noticed earlier that its radio emission pattern was elongated, but when Jennison and Das Gupta zeroed in on it they found that the apparent elongation was actually two separate lobes. Cygnus A was in reality two sources; the radio emission was coming from two different regions of space. When these regions were compared to the optical galaxy it was found that the galaxy was in the center; the lobes were symmetric on either side. Astronomers had little idea what caused this or what it meant.

It is important to understand how radio plots are made by astronomers, so let's take a moment to consider it. The radio intensity around a source is measured, and contours of equal intensity are plotted. This means that the plot of a source will look like a series of concentric rings, with the highest intensity at the center. If you have ever done any mountain climbing or hiking you've likely seen similar rings on a map. These rings, or contours, tell us where the altitude is the same. A mountain will therefore appear as a series of concentric contours, with the highest point in the center.

What Jennison and Das Gupta found was that there was a series of contours on either side of Cygnus A. Techniques were

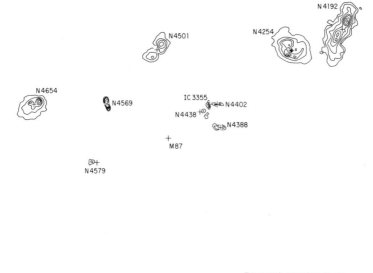

Total HI in the Virgo Cluster

Radio contours around radio sources. [National Radio Astronomy Observatory (NRAO)]

still relatively primitive at this stage, so they could do little more than tell that it was a double source. Nevertheless, a double source did not seem to be consistent with two galaxies in collision. The two sources were well outside the optical galaxy, and there seemed to be no way that two colliding galaxies would give off radiation in this strange way.

A much more detailed map of the lobes was obtained in 1956 by R. Hanbury Brown and his colleagues at Jodrell Bank. Brown used the Mark I as part of an interferometer; it was linked to another smaller radio telescope about 12 miles away. To complete the interferometer they needed readings between the two

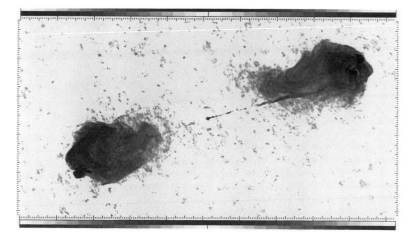

Radiograph of Cygnus A showing lobes. Cygnus A is the small object at the center. [National Radio Astronomy Observatory (NRAO) and observers R. A. Perley, J. W. Dreher, and J. J. Cowan]

larger telescopes so Henry Palmer drove across country between the two making readings wherever he could. The result was an impressive picture of Cygnus A. There were definitely lobes on either side of the visible galaxy, and overall the system was huge—320,000 light-years across. This made it one of the largest known systems in the universe. It appeared as if gas was somehow being expelled from the central galaxy, creating the lobes. But what caused them? And why were they so highly energetic?

About this time the coordinates of another radio source arrived on the desk of Rudolph Minkowski at Palomar. It had all the appearance of being another source like Cygnus A, and Minkowski immediately looked up the position on the newly printed Palomar–National Geographic Sky Survey plates. There was nothing visible at the position (it was in the constellation Böotes). He therefore took an exposure with the 200-inch reflector, and found an exceedingly dim cluster of galaxies. Minkowski then took the spectrum of the brightest of the galaxies and was amazed by its incredibly large red shift. It was much greater than that of Cygnus A, indicating that it was 10 times farther out. Later measurements showed that, like Cygnus A, it also had lobes on either side of it.

EXPLODING GALAXIES?

With the discovery of lobes and the discovery that not all of the radio sources appeared to be double galaxies, the idea of colliding galaxies quickly died away. It seemed unlikely that all the energy was being generated by two galaxies in collision. In fact it was virtually impossible. The only alternative was an exploding galaxy. The evidence that some sort of explosion was going on was indeed overwhelming. Photographic evidence coupled with the fact that the radiation was synchrotron radiation made this an attractive hypothesis. But astronomers still had to solve the problem of what was causing the explosions.

Soon after the discovery of Cygnus A another radio galaxy was discovered in the south by Australian radio astronomers. It was in the constellation Centaurus (and was soon called Centaurus A). It was huge, its size primarily a result of its proximity; it was only about 15 million light-years away from Earth. Like Cygnus A it also had lobes, two large ones about a million light-years out from the galaxy. But, interestingly, it also had two much closer; they were only 16,000 light-years from the nucleus. Also, like Cygnus A, it appeared to be two galaxies in collision, but it was later seen that the double appearance was due to a band of dust across its center.

Centaurus A is a giant elliptical galaxy, containing about three times as many stars as our Milky Way. Like most radio sources it is also a strong source of X rays and other radiation. Most of its energy comes from its nucleus, but a considerable amount comes from its two pair of lobes. These lobes are likely gaseous—perhaps gas ejected from the nucleus.

Interestingly, most of the strong radio sources are ellipticals like Centaurus A and Cygnus A. We will see later, though, that Centaurus A may indeed be two galaxies in collision—an elliptical colliding with a smaller spiral.

Another strong radio source is the Messier object M87, a giant elliptical in the constellation Virgo. The most intense radiation comes from the center, with a smaller amount coming from the halo. Of particular interest, however, is a jet that is emanating from its core; it is also a strong source of radiation. It looks as if it is material that has been ejected from the core. The explosion required to produce such a jet, however, would have to be extremely powerful. A closer look at this jet shows that it consists of several highly condensed regions.

Explosive power is also evident in a strong source in Perseus, called Perseus A. It has long filaments of gas, hundreds of thousands of light-years long, emanating from its core. Doppler measurements show that the gas has a speed of about 3000 km/sec. Its filamentary structure makes it look like the Crab Nebula.

So far, all the sources I've talked about are strong. But it seems reasonable that if the energy is, indeed, being produced by an explosion, that a few of the galaxies would be much weaker sources. After all, the explosion may start slowly, and of course, it eventually has to die down. Do we have any intermediate cases? Indeed we do. One is in the constellation Ursa Major and it is known as M82.

M82

In 1961 Roger Lynds of Palomar began trying to get a photograph of the radio source 3C 231. The radio image was fuzzy so its position wasn't accurately known, but it had been narrowed down to a small box. The first photograph Lynds took showed a cluster of galaxies, the brightest of which was the spiral M81. He assumed it was the source. But with a little more work he realized it wasn't M81, but the smaller nearby galaxy M82 that was

M82. Note filaments extending out from the core. (Hale Observatories)

the source. M82 was an odd-looking galaxy; Hubble had classified it as an irregular, but it was difficult to determine how it was being viewed. If we were seeing it near edge-on, it could be a spiral. (This was later shown to be the case.) For the most part its spectrum was ordinary, although some emission lines were visible. And it also showed that there were a lot of young stars present. It was therefore a relatively young galaxy.

Its spectrum also showed that it was nearby—only about 10 million light-years away. Once M82 was identified as the radio source, Allan Sandage took several photographs of it with the 200-inch reflector. Instead of taking them in blue light, which was the usual case, he used a red filter. And, to his surprise, huge filaments were seen emanating from the core. Some of them were 1000 light-years long. Some sort of explosion was obviously going on in the core. Doppler measurements indicated that the filaments had speeds of 1000 km/sec. This meant that a tremendous amount of material was being ejected—an amount equal to several million suns.

Although photographs strongly suggested that the nucleus of M82 was in the process of exploding, it was not a particularly strong radio source. It was much weaker than either Cygnus A or Centaurus A. Alan Solinger, Tom Markert, and Philip Morrison of MIT therefore decided to take a closer look at the strange core. Using special near infrared plates that could penetrate the dust they were able to see deep into it. If the nucleus really was exploding, there should have been a bright, dense region in the center—the "engine" that produced the power. To their surprise, though, they didn't find a bright nucleus; they found clusters of exceedingly bright stars—huge young stars. The core was composed of large numbers of young stars, but there were also young stars in the outer regions. These bright stars were scattered across a region of about 1000 light-years.

M82 wasn't so strange, after all. In most respects it was like an ordinary galaxy except for its radio emission (which was quite weak, anyway) and the visible filaments. The bright stars at the center, however, were an enigma. A study of the polarization of the central region was also made, and it was inconsistent with an explosion. Maybe M82 wasn't exploding after all. A violent explosion was also inconsistent with its radio emission. The emission could be accounted for by roughly a thousand supernovae and, if the central region consisted of a large num-

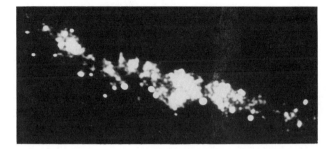

Radio map of M82 showing hot spots and jetlike features. [National Radio Astronomy Observatory (NRAO) and observers P. P. Kronberg, P. L. Biermann, and F. R. Schwab]

ber of young stars, there likely were a thousand supernova remnants.

But what caused the sudden surge of new star making near the core? Halton Arp of Palomar took a long exposure of the region and found a faint cloud of gas surrounding both M81 and M82. The pieces of the puzzle started to fall together. It looked as if a huge cloud of gas from M81 had struck M82. In fact, astronomers were able to estimate when it took place; they arrived at a figure of about 100 million years. If such a cloud did strike M82, much of it would have fallen into the core with a resulting burst of stellar activity. Large numbers of new stars would have been born, and within a short time the largest would have become supernovae.

But what about the visible evidence of an explosion—the filaments? Solinger and his group believe that this is mostly dust that is being carried outward on gas clouds that are moving at relatively low speeds.

The weak radio emission from M82 is now believed to be coming from the supernovae, but for most active galaxies this is not the case. They have dense, compact cores that power the galaxy. Oddly enough, it had been known for years that certain types of galaxies had dense, compact cores.

SEYFERTS AND N GALAXIES

In 1943 a young postdoctoral student at Mount Wilson Observatory, Carl Seyfert, made an interesting discovery. He noticed that short exposures of a number of spiral galaxies showed an extremely intense center—like a bright star. Yet when a long exposure was made, it looked like an ordinary galaxy. Seyfert identified about a dozen of these galaxies, and they are now named after him. Unfortunately he didn't live to see the excitement that his discovery created. A few years after his discovery he was killed in an automobile accident.

The cores of Seyfert galaxies are so bright that they outshine ordinary galaxies 100 times. Furthermore, their spectra are different from ordinary galaxies: they contain emission lines indicating activity near the core. In addition, some of the emission lines are broad, indicating that clouds of gas are in rapid motion near the core. But, interestingly, not all the emission lines are broad. This has puzzled astronomers. It seems to indicate that there are more than one group of gas clouds in rapid motion, some of which produce broad lines, others of which do not.

Closely related to Seyferts are another class of radio object called N galaxies. Seyfert galaxies are, in fact, a subclass of N galaxies, distinguished by the emission lines in their spectrum. As in the case of Seyferts, N galaxies have bright cores. A short exposure of, say, one minute shows a starlike object with a slight bit of fuzz around it, whereas a longer exposure of 10 minutes shows the surrounding arms of the galaxy. Some N galaxies appear disturbed in optical photographs; some have visible filaments, others have jets. Yet, many N galaxies appear completely normal in photographs.

The discovery of Seyferts and the apparent exploding core of M82 were the first indications that there was a class of galaxies quite different from ordinary galaxies—galaxies with extremely active cores. We now refer to them as *active* galaxies. One of the interesting things about active galaxies is that few are nearby. If we examine the galaxies around us we find that all of

Seyfert galaxy. (National Optical Astronomy Observatories, Tucson, Arizona)

them, with the exception of a few, such as M82, are ordinary galaxies. We have to go far into extragalactic space before we find active galaxies. In fact, we find that as we look farther and farther into space galaxies become increasingly active. But, of course, because of the finite speed of light, when we look into

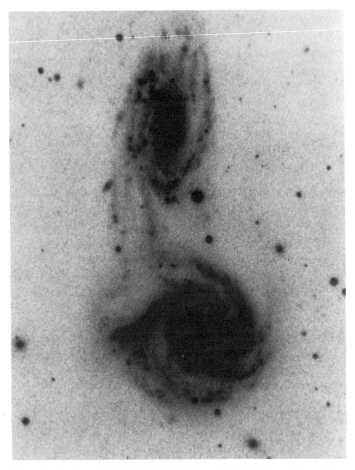

Peculiar galaxies from Arp's Atlas *(printed in negative). (Palomar Observatory and Halton Arp)*

space we are actually looking back in time. Distant galaxies are therefore seen as they were billions of years ago. A galaxy 2 billion light-years away is seen as it was 2 billion years ago; similarly one 5 billion light-years away is seen as it was 5 billion

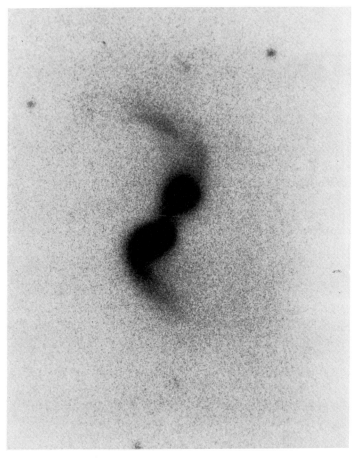

Peculiar galaxy from Arp's Atlas. *(Palomar Observatory and Halton Arp)*

years ago. Does this mean that galaxies, when first formed, are active, then, as they age, lose activity? There is some speculation that this might be the case, but at the present time most astronomers do not accept it.

PECULIAR GALAXIES

Active galaxies have become a challenge to astronomers. What is the "engine" that powers them? And what is the relationship between the various types of active and ordinary galaxies? Why are some active, others not? Many ideas have been suggested and rejected over the years, and some progress has been made. Yet many questions remain.

There are so many strange-looking, or peculiar objects in the universe that an astronomer named Halton Arp decided in the early 1960s to make an atlas of them. Arp did his undergraduate work at Harvard, and obtained his Ph.D. from Caltech. After working on variable stars and globular clusters for a while he became intrigued with the large number of peculiar galaxies that were visible with the 200-inch telescope. Some of these galaxies appeared to be exploding, others had jets emanating from them. Still others appeared to be galaxies in collision. He also found an object that was very far away, according to its redshift, yet it appeared to be connected by a "bridge" to a much nearer object. His atlas is now referred to as the *Atlas of Peculiar Objects*. Initially, it contained only objects from the northern hemisphere, but he later extended it to include objects in the southern hemisphere.

Arp has found so many peculiar objects he has become convinced that the some of the basic laws of physics, as we now know them, need modification, or perhaps, replacement. But most astronomers have not heeded his call; they consider him a maverick who has gone too far with his speculation.

CHAPTER 5

Cosmic Jets and Galactic Dynamos

The discovery of diffuse gaseous radio lobes on either side of active galaxies took astronomers by surprise. The lobes were huge—much larger than the galaxy itself. And their distance from the galaxy, in some cases, was up to 10 million light-years. How were they connected with the galaxy? It was reasonable to assume that they were being produced by the galaxy. But how? Astronomers were stumped.

Then they discovered that the electric fields of the radio emissions were oriented in a particular direction in space. This meant that the radiation had to be synchrotron radiation, in other words, radiation that is produced by charged particles spiraling along magnetic field lines. The lobes, therefore, had to be filled with electrons, protons, and magnetic fields.

A further surprise was the discovery of "hot spots" in the clouds. These were regions where the intensity of radiation was extremely high. And, strangely, they were always on the side of the lobes away from the galaxy. Why? What was causing them? Their energy, along with the other energy of the lobes, was enormous—equivalent to the energy released in the annihilation of millions of stars. It was unlikely that the lobes themselves were producing this energy. Somehow it had to be generated in the galaxy and transported to the lobes. But how?

Finally, a connection was made. Long narrow jets were discovered extending from the galaxy out to the lobes. Tracing the jets back they were found to originate at the nucleus of the

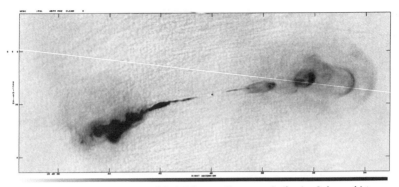

Radiograph of Hercules A, one of the brightest radio sources in the sky. Lobes and jets are easily seen. Galaxy is at the center. [National Radio Astronomy Observatory (NRAO) and observers J. W. Dreher and E. D. Feigelson]

galaxy. They were narrow beams of gas shot out from the core of the galaxy. Hot gas from deep within the core was being formed into a long jet, hundreds of thousands of light-years long. At the ends of the jet the gas was spreading out to produce the lobes.

One of the first jets to be studied was associated with the galaxy NGC 6251. It extended out 400,000 light-years from the galaxy, and was exceedingly narrow and straight. Tracing it back to its source, astronomers found a small region of intense radiation at the center of the galaxy. And right next to the center was a much smaller jet. The source had to be incredibly powerful to produce such jets, yet it was tiny.

NGC 6251 appeared to have only a single main jet, but many radio sources have two. For example, 3C 449 has long jets emanating from either side. Interestingly, each of them are bent, but the bends are the mirror image of one another. In other cases, however, such as NGC 1265, the bends are in the same direction. And in 3C 75 there are two sets of jets: four enormous jets blast out of the nucleus, all with bends.

Richard Perley and his group at the Very Large Array (VLA) have developed a technique that allows them to obtain photographlike pictures, called radiographs, of the jets and lobes. They use the computer to obtain a gray scale for each region of the lobe and jet, then print out the results (see figures).

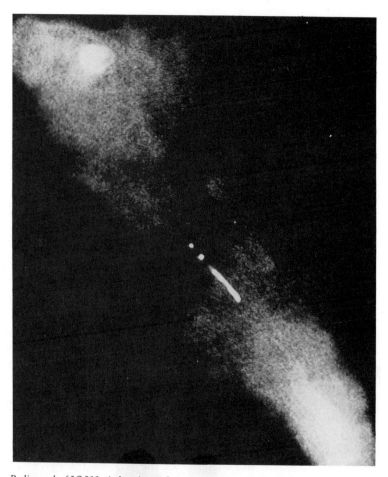

Radiograph of 3C 219. A short jet can be seen near the center. [National Radio Astronomy Observatory (NRAO) and observers A. H. Bridle and R. A. Perley]

RELATION BETWEEN JETS AND THE NUCLEUS

By the late 1970s astronomers had established that the jets were conduits from the powerful "engine" in the core out to the lobes. But many problems remained. It was difficult to deter-

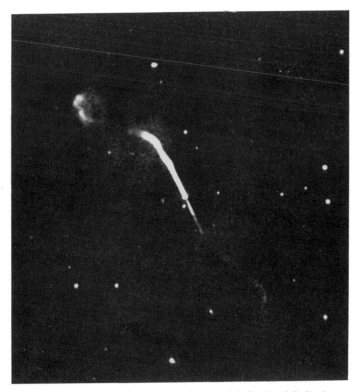

Closeup of jet. A weak counterjet is seen to the upper left. [National Radio Astronomy Observatory (NRAO) and observers A. H. Bridle and R. A. Perley]

mine the speed of the material in the jets, as there was no apparent motion. Estimates, however, ranged from a few hundred to perhaps 10,000 miles/sec, and this presented a serious problem. Since the distance to the lobes was typically half a million light-years, there was no way that radio electrons spiraling outward along the magnetic field lines had a lifetime long enough to reach the lobes. It would take millions of years for them to get there, and the lifetime of the electrons couldn't be more than a few hundred years. Something had to be pumping energy into the beam as it forced its way outward. Furthermore, energy had to be supplied continuously from the core. The jet couldn't have been caused by a single explosion. In addition, the source had to remain pointed in the same direction for thousands and, in some cases, for hundreds of thousands of years, as the jets were long and, in many cases, straight.

Another enigma was the diameter of the jets. They were so narrow it appeared as if they had been created by gas forced through a small nozzle, in the same way that gas in a welder's torch is forced through a small nozzle. But how was such a nozzle created? You wouldn't expect to find a small circular opening near the center of a galaxy. One of the most logical solutions was that it was caused by whirling gas. A cloud of gas in the form of a "whirlpool" would have a hole in the center.

One of the best examples we have of jets here on Earth are geysers such as Old Faithful in Yellowstone. In this case a cavern beneath the Earth contains water. As the water heats, steam pressure builds up until it can no longer be contained. It then forces its way up through a small nozzle near the surface and emerges at high speed, creating a jet of hot water and steam that rises several hundred feet above the earth.

It is reasonable to assume that in the case of radio galaxies we have a similar phenomenon—a galactic geyser. In this case we would have superheated gas near the core. Around it would be a swirling cloud of cooler gas. If this cloud of cooler gas were in the form of a fat doughnut with a tiny hole in the center, and if the hot gas were entirely surrounded by this doughnut, it would have two openings through which to escape—one on either side of it.

When the pressure built up sufficiently the hot gas at the center would surge out of the two nozzles. Pushing its way through the interstellar medium, it would continue at high speed until it encountered intergalactic gas. And, for a while, it would continue pushing this gas out of the way, but eventually the intergalactic gas would take its toll and the gas in the jet would begin to decelerate. Since it is made up mostly of charged particles, it would radiate, and radio waves would be given off. The greatest deceleration would occur at the ends of the lobes—creating the hot spots. Charged particles would be splashed around creating the lobes and keeping them energized.

The model seems reasonable, but it leaves many questions unanswered. Where does the fuel—the plasma—come from? What produces it? Why is it so energetic? There has to be some sort of "engine" in the core producing all the energy. What is it like? Before we answer these questions, let's look at one of the jets in detail.

JETS IN M87

One of the best known and most extensively studied jets is in the giant elliptical galaxy M87 in the Virgo cluster. It was discovered in 1918 by H. D. Curtis of Lick Observatory. He had no idea what it was. At the time astronomers were still arguing about whether the nebulous white clouds were island universes of stars or not, so they were still not even certain that M87 was a galaxy. For years it therefore remained nothing more than a curiosity. But when radio astronomy came into its own after World War II astronomers soon discovered that M87 was a powerful radio source.

Then in 1954 Baade and Minkowski of Mount Wilson Observatory showed that much of the radio emission was coming from the jet. Others began to get interested. Two years later Baade and Minkowski showed that the optical emission from the jet was polarized. Then radio astronomers noticed that the radio emission was also polarized. This meant the jet had to be composed of charged particles spiraling along magnetic field lines. Finally, in 1966 X-ray astronomers took a look at the jet and found that it was also a strong source of X rays. (It was the first identified extragalactic X-ray source.)

The jet in M87 is seen only in short exposures; the light from the galaxy itself obliterates it in long exposures. It is about

A short exposure of M87 showing jet. (Lick Observatory, University of California, Santa Cruz, Calif. 95064)

6500 light-years long, which is short compared to some of the jets I talked about earlier. But we still have the same problem. In fact, it's more serious in this case. The radiation is being caused by energetic electrons, yet radio energy electrons have lifetimes of only a few hundred years. The real problem, however, is that the jet is also a strong source of X rays, and if the electrons were energetic enough to produce X rays they would last only a few days. Something has to be re-energizing the electrons.

Another problem relates to the jet itself. Why is there only one jet? The explanation we talked about earlier assumes that there have to be two oppositely directed jets. We will see later that most astronomers believe that there are two jets but that we are only seeing the one closest to us. In fact, we have photographic evidence of a second jet. Halton Arp made a long exposure that showed an exceedingly dim jet on the other side.

Another interesting aspect of the jet is its structure. A detailed look at it using a charge-coupled device (CCD) showed that it has a "lumpy" structure. Indeed, it appears to have knots in it. Astronomers at the VLA, however, have recently obtained high resolution pictures of the jet and find that it is not as lumpy as first indicated. These pictures show a jet that is bright, with dim regions. They also suggest that the emission may be coming from the surface of the jet rather than the inside; in other words, the jet appears to be "lit up" by the layer on the outside. And in places this layer is dim, giving the appearance of a knot.

The most intriguing question, though, is: What is causing the jet? Obviously, as in the case of other active galaxies, an "engine" must exist at the core. In 1978 two teams of astrophysicists at Caltech decided to take a closer look at this engine. One was headed by Peter Young, the other by Wallace Sargent. Young's group measured the light intensity across the jet. Since the light comes from stars, its distribution would give an estimate of the number, or density, of stars within the galaxy. The group under Sargent took spectroscope measurements at various positions across it. Any given spectrum would be made up of the light of thousands of individual stars. Nevertheless, it would give information about the average speed of the stars at various distances from the center.

The results from the two groups were astounding. First of all, M87 was found to be excessively bright near the center, compared to other ellipticals, indicating a high density of stars.

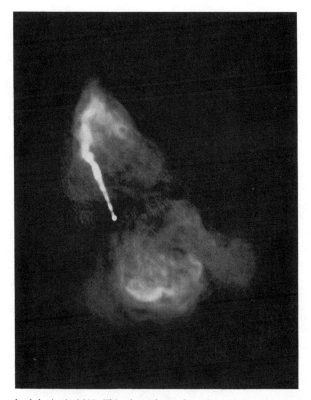

*Radiograph of the jet in M87. This photo shows the radio emission; the previous photo
shows the optical emission. [National Radio Astronomy Observatory (NRAO) and observers F. N. Owen and D. C. Hines]*

Second, the spectroscopic measurements showed a large spread
in stellar velocities across the disk of the galaxy. The velocities
near the center were incredibly high. The only way these things
could be explained was if there was a huge dark mass in the
center. In fact, the object had to have a mass of five billion suns.
Yet it had to be small.

A model began to emerge. It seemed that a massive engine

Jets emanating from spiral galaxy.

at the center of the galaxy was being fueled by nearby gas, and this gas had built up enough pressure to produce an extremely energetic jet.

But how was the jet being fueled? The appearance of M87 and the galaxies around it may be a clue. M87 is at the center of a huge cluster of galaxies; furthermore, it is the largest of all the galaxies. In addition, nearby galaxies appear to have been stripped of much of their outer gas. Is it possible that M87 cannibalized the outer gas from these galaxies? Is this the fuel that produces its energy? Many astronomers believe that this may be the case.

BLACK HOLES

Still, we have the problem of the "engine." As we saw earlier, it has to be extremely massive, yet small. Variations in the brightness of active galaxies, in fact, indicate that it cannot be much larger than the solar system. Does a small, massive object that is capable of producing vast amounts of energy exist? Indeed, it does: the black hole.

To understand black holes we have to begin with stars. Like people, stars go through a life cycle. They are born, live a certain amount of time, depending on their mass, then they die. They are born when huge gas clouds gravitationally collapse, heating their centers to millions of degrees. When the core gets to a temperature of about 15 million degrees, nuclear reactions are triggered. Energy then begins to flow out, creating an outward force which eventually becomes strong enough to balance the strong inward pull of gravity. The two forces balance and the star reaches equilibrium. For billions of years, or millions in the case of very massive stars, the star peacefully consumes its fuel. Its fuel at this stage is hydrogen, but later it may burn heavier elements such as helium, carbon, and oxygen, depending on its mass.

In time, though, the star will begin to run out of fuel, and as it does, the outward force will diminish. Eventually it will be overcome by gravity and will begin to collapse inward on itself. If the star is about the mass of the sun, the collapse will take millions of years. But if it is supermassive, the collapse will occur rapidly—in seconds, or even less. In fact, if the star has a mass greater than about 8 solar masses, it will collapse in a tiny fraction of a second. It will become a black hole.

The first scientists to study this collapse in detail were Robert Oppenheimer and his student Hartland Snyder. In 1939 they found that if a star was sufficiently massive, once the collapse started it was impossible to stop. It continued until all the mass had collapsed to a point. Yet, strangely, as the collapsing star passed through a certain radius, called the gravitational radius, it left a mysterious black spherical surface in space. This surface is now referred to as the event horizon. The only requirement for such a collapse is that the final mass be slightly over 3 solar masses.

One of the most interesting parts of the black hole is the event horizon. It is, in essence, a one-way membrane. It is not solid so you could pass right through it, and although it looks like a black balloon, it won't pop as you pass through it. In fact, except for a sensation of tremendous force on your body, you wouldn't even know you were passing it. But once you were inside there would be no escape. It takes a velocity greater than that of light to get back out, and according to Einstein's theory of relativity, this is impossible. The uppermost speed of any massive object in the universe is the speed of light.

The radius of the event horizon depends on the mass of the star that collapsed to create it. The more massive it is, the greater the radius. This means that as stars and space debris fall into a black hole they cause it to grow. Furthermore, if two black holes collide, they merge, creating a much larger black hole.

Most black holes, according to our current models, also spin. And if you were to come close to a spinning black hole you would be pulled around in the direction of the spin. This type of black hole is called a Kerr black hole.

The black holes I've talked about so far are what are called stellar-collapse black holes. They occur when stars collapse. But individual stars don't have enough mass to create the kind of black hole we need for active galaxies (the dark mass in M87 is about 5 billion solar masses). Do even more massive black holes exist? According to Stephen Hawking of Cambridge University they do. He pointed out in the early 1970s that supermassive black holes may have been created in the big bang explosion that gave rise to the universe. According to Hawking, if the big bang explosion was inhomogeneous (nonuniform), pockets of matter would have been compressed into black holes. And there is, indeed, strong evidence that the big bang was inhomogeneous. If it wasn't, galaxies wouldn't have been created, and we see galaxies all around us—including our own. To distinguish this type of black hole from the stellar-collapse variety we refer to them as primordial black holes.

Stars do not vary greatly in mass, and therefore stellar-collapse black holes are all roughly the same size—a few miles across. Primordial black holes, on the other hand, range tremendously—all the way from tiny, microscopic-sized ones up to those with the mass of a galaxy. Some astronomers believe that primordial black holes served as "seeds" for galaxies. If so, all galaxies would have supermassive black holes at their center. Active galaxies would be energetic because they still have a source of fuel around the black hole; the black hole is gorging itself on stars that are spiraling into it. Ordinary galaxies, on the other hand, are radio-quiet because, even though they have a black hole at their center, there are no longer any stars close to it. In short, the black hole's fuel supply has been depleted.

If active galaxies do, indeed, have giant black holes at their center, many of their properties can be explained. Their engine has to be small and exceedingly massive, and supermassive black holes have these properties. Furthermore, they are capable of energizing the jets and the lobes.

BLACK HOLE MODELS

Earlier we saw that one of the best models of the jets assumed that they resulted when a cool cloud of spinning gas surrounded hotter gas. If the gas was in the form of a whirlpool with a small hole in the center, the inner gas could be trapped inside the center and ejected out through two holes in opposite directions.

But what would cause the cloud to spin, and what would supply the energy to heat the inner gas and force it out the nozzle? This, of course, has to be supplied by the small but powerful engine at the center. Our discussion of black holes in the last section leads us to believe that the best bet for this engine is a supermassive black hole. A black hole with a billion solar masses would only be slightly larger than the solar system. A spinning black hole of this size and mass could easily cause a huge gas cloud to spin, and it could supply the required energy.

The first black hole model of active galaxies was proposed in 1968 by the British astronomer Donald Lynden-Bell while working at Caltech. He assumed that the stars, gas, and so on around the giant black hole formed a huge accretion disk around it, much like that of Saturn, only larger. Important modifications to Lynden-Bell's model were made by Roger Blandford of Caltech and Richard Lovelace of Cornell. They assumed the black hole was spinning, and that it had a magnetic field. Both are reasonable assumptions in that we know that our galaxy, and for that matter, all galaxies, spin. Furthermore, all galaxies have magnetic fields.

Spinning black holes drag everything that comes close to them around in the direction of spin. Because of this they eventually are surrounded by a giant accretion disk. In this disk the inner material will move faster, just as it does in the case of Saturn's ring, and this in turn will create a viscous friction that will cause the material to gradually move inward. And as it moves it will carry the magnetic field lines with it, until finally the magnetic field near the black hole becomes exceedingly strong.

Furthermore, as the gas spirals into the region near the black hole it will be stretched and compressed, causing it to heat. The gas closest to the black hole will therefore be the hottest—its temperature eventually reaching a billion degrees. Surrounding it is a cooler spinning gas cloud. Because the inner gas is much hotter, it will eventually force its way out through the funnels in the surrounding cloud. And as it does it will be channeled into two high-speed, narrow jets.

Mitchell Begelman.

Several astronomers are now trying to understand the details of how this model works. One of them is Mitchell Begelman of the University of Colorado. Upon graduation from the Bronx High School of Science in New York, Begelman went to Harvard where he obtained a bachelor and master's degrees in physics. From there he went to Cambridge University in England to work on a Ph.D. His thesis, which was done under Martin Rees, was on the theory of accretion onto black holes.

I asked him how he first got interested in astronomy. "My initial interest goes back a long way," he said. "It started when I was about five or six. I recall becoming interested in solar system exploration first . . . about the time Sputnik was launched. There was a lot about Sputnik in the news and it impressed me. Furthermore, my parents encouraged my interest. They got me a small telesope when I was six, so I was an amateur astronomer right up to the time I went to university. When I was in high school I was very active in some of the New York astronomical societies. I observed all the objects in Messier's catalog with my telescope—most from the roof of our apartment building in the Bronx."

One of the problems that Begelman has worked on is the form of the magnetic field lines near a spinning black hole. "There has been a lot of work on magnetic effects which might generate jets," he said. "And I think most people are now convinced they are the best models." Using an interesting analogy, he described to me what would happen to these lines. "Suppose you have a straight magnetic field line pointing directly out of the center and assume the gas is like a bead that slides along this line," he said. "To a good approximation the bead has to stay on the field . . . it can't move across it. This means that the field lines have to move where the bead moves and vice versa. When you start this thing spinning with the magnetic field line pointed radially outward there will be a centrifugal force that will fling the bead outward. Eventually, as it moves outward, its speed will get close to that of light. But when you move something this fast its inertia [mass] goes up. So you find that you have enough force to bend the field lines back, and they start to spiral around the black hole."

The lines start to spiral around the black hole because they cannot travel at the speed of light. They will start to wind up, but eventually will become "overwound." And, as a result, a tremendous tension will develop that will have to be relieved. The magnetic field lines near the black hole, which normally rotate at the same speed as the black hole, will therefore begin to slip. And this slipping will generate a powerful electric field. The overall effect is like that of a huge dynamo, with electrons being propelled outward along the electric field lines.

One of Begelman's other interests is the fueling of black holes. He is attempting to answer the question: Where does the gas come from that is shot out in the jets? "We've been studying the different mechanisms by which gas could be forced to flow into the engine," he said. "It seems likely that the gas that is flowing toward the center would go through a region where it would become unstable and form a large number of stars. In some cases bursts of star formation appear to be associated with activity in the nucleus."

I asked him if he thought "cannibalism" was an important in fueling. (By this I meant "gobbling up" nearby galaxies.) "I think objects like 3C 75 certainly suggest that something like that is going on. But I suspect that a wide range of disturbances in a galaxy could cause materials to flow into the center and fuel the activity. The source of fuel might be stars from the galaxy itself . . . and it doesn't have to be ones close to the black hole. If you can figure out a way to get the gas into the center . . . which means you have to get rid of its angular momentum [spin], then there's enough gas in almost any normal spiral to turn it into an active galaxy."

Another astronomer who is working on jets is Sterl Phinney of Caltech. He is primarily interested in the composition of the jets. "An important question in relation to the jets," he said. "is whether they are formed of electrons, positrons [positive electrons], or electrons and protons." Phinney is attempting to answer this question by making models of jets using these particles, and determining whether they are stable or not. He has found that there are difficulties with the models he has devised so far, but he is confident that the problems can be overcome.

THE MYSTERY OF SINGLE JETS

We saw earlier that some active galaxies appear to have only one jet. But the model I talked about earlier indicates there should be two jets. One way of explaining this is to assume that there are two jets in all cases, but that in some cases one of the jets is not visible. To explain this, let's consider what happens to a source of radiation that is traveling at high speed. It is well known that it will tend to focus its radiation in the direction in which it is traveling. And the faster the source travels, the sharper it focuses its radiation. At speeds close to that of light the radiation is focused into a narrow, sharp beam. Because of this, radiation that is traveling toward you from sources that have a speed near that of light will appear much brighter than they would if they were stationary—up to 1000 times brighter, depending on their exact speed. On the other hand, if you are not in the direction the beam is pointed (or close) it would be virtually invisible.

Using this as a guide we can explain what various sources would look like. We assume in all cases that the radiation is beamed out of the source in two opposite directions. If one of the two jets was pointed at a small angle to your direction you would see one of them as a bright jet. The far jet would be invisible. If it were exactly in your direction it would be narrow, and particularly bright, and there might be a slight halo around it. Finally, if the two jets were perpendicular to your direction, neither would be seen.

Strong evidence for this point of view was obtained by Bob Liang and his colleagues at Greenwich Observatory recently. They showed by analyzing the polarization of a beam that when we see a single jet it has to be on the side closest to us.

JETS WITH THE BENDS

We saw earlier that jets connect the engine at the center of the galaxy to the lobes. But the jets linking these regions are not necessarily straight. Many jets, in fact, have bends: 3C 449, for example, has mirror-image bends. NGC 1265 has an interesting radio tail, and 3C 75 has several bends in its two sets of jets.

What causes these bends? There are three known explanations. The first is related to the stability of the beam. It is traveling at supersonic speeds, and as it moves outward it pushes the gas in front of it out of the way. This gas may not be uniform, and if an inhomogeneity is encountered it will affect the jet beam. If the pressure, for example, is decreased on one side it will cause the jet to change direction slightly.

The second method relates to the the gas that permeates our galaxy—called the interstellar medium. It can act like a wind

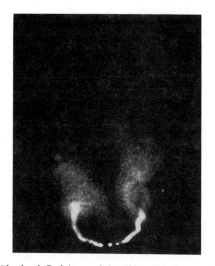

Radiograph of jet with a bend. Both jets are being "blown" in the same direction. [National Radio Astronomy Observatory (NRAO) and observers C. P. O'Dea and F. N. Owen]

and blow the jets sideways. In this case the bends on the two jets will be in the same direction, resulting in what are called radio tails. You would get the same effect with twin chimneys. A breeze would blow the two smoke columns into a single one.

Third, a second galaxy may cause the bend. A spinning black hole, for example, might be precessing, just as a spinning top here on Earth precesses (its spin axis traces out a cone). The jet would then leave the zigzag pattern of the precession. Or it might be pulled in the direction of a second galaxy by the galaxy's gravitational pull. In some cases the second galaxy might not be visible so we would not know what was causing the bend.

SUPERLUMINAL SPEEDS

In 1971 it was discovered that radio-emitting material near the center of some of the active galaxies appeared to be moving at

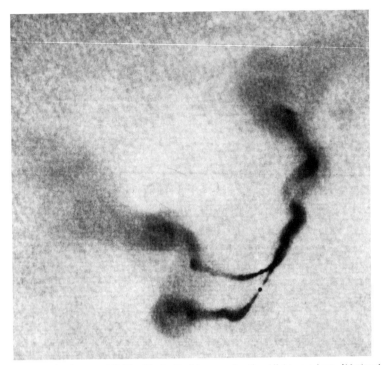

Radiograph of source 3C 75 with double jets on each side. All jets are bent. [National Radio Astronomy Observatory (NRAO) and observers F. N. Owen, C. P. O'Dea, and M. Inoue]

speeds greater than that of light—superluminal speeds. Astronomers were amazed. This was impossible; nothing can move faster than the speed of light according to relativity theory. Yet near the center of these galaxies there were indications of speeds up to 20 times that of light.

The problem, as it turned out, was related to the relativistic changes that occur near the speed of light. It was, in effect, just an illusion caused by the way we view it. If one of the beams is in our general direction, it will be greatly intensified due to the effect I described earlier. The faster-than-light motion results from the source traveling toward us at close to the speed of light. The radio signal travels at the speed of light and the source at slightly less. This means that two bursts of radiation will appear to be separated by much less than they really are. They might appear to be separated by a week, when in reality they are separated by several months. Without taking the motion of the source into consideration we calculate speeds that are greater than the speed of light. Once we understand what is going on, however, and make corrections for it, we get speeds less than the speed of light.

OTHER MODELS

While the black hole model is accepted by most astronomers, some are skeptical. They feel that the same predictions can be made using a far less exotic engine. Dense regions of exploding stars, or groups of tiny stars such as white dwarfs or neutron stars would do the same job, they claim. Of course the density would have to be exceedingly high—a million times that of our region of space.

Begelman and many other astronomers don't agree. "It can't just be a loose collection of supernovae," said Begelman. "We've seen recently in X-ray observations that the variability of the X rays is on the scale of a few months. You see the luminosity fluctuating by a factor of two in less than an hour. It would be almost impossible to explain this with a collection of stellar objects. You can't really prove the engine has to be a black hole, but if it isn't, it has to be something that is very close. You have to have an object that is so compact that its escape velocity is within a few percent of the speed of light. And so far no one has come up with a stellar collection model that is stable enough." He paused, then said, "I think the most economical hypothesis by far is the black hole one."

Richard Perley of the VLA says, "I think the evidence is very much in favor of the black hole model. This doesn't mean the problems are all worked out, but they have the energy, and mechanisms have been worked out that show considerable promise of being able to produce all observed signatures. The possibility that huge black holes may exist at the cores of active galaxies is, indeed, an exciting one. And within the next few years, with new instrumentation and better observing techniques, astronomers will no doubt find out for sure whether or not this is the case.

CHAPTER 6

Quasars

It might be hard to believe that something even more powerful and more energetic than a radio galaxy could exist. But it does. In the early 1960s extremely energetic objects that we now call quasars were discovered.

The discovery came about when astronomers began trying to identify some of the radio sources in the 3C catalog. Some of them were quite easy to associate with known optical objects; many, in fact, were galaxies. In most cases these galaxies appeared to be abnormal in some way: exploding or colliding. But some of the stronger sources could not be identified. Photographs of the regions showed only a field of stars and distant galaxies. Because of the low resolution of the radio telescopes it was extremely difficult to determine which of the objects in the photograph was the source of the radio waves.

In the late 1950s Tom Matthews, a radio astronomer at Caltech, selected 10 of these objects and decided to make a serious effort to identify them. The going wasn't easy, but within a short time he managed to get the coordinates of one known as 3C 48. Matthews passed its position on to Allan Sandage, who photographed the region with the 200-inch Palomar reflector. The only object at the position was a dim bluish star. It didn't look spectacular in any way, which made things all the more puzzling. The reason astronomers were bewildered was that until that time stars were not known to be radio sources. Our sun was a source of radio waves, but we could detect them

Jesse Greenstein.

only because it was so close. Other than the sun, though, the only known radio sources were galaxies, or supernova remnants such as the Crab nebula.

Sandage then took the spectrum of the object. It was even more puzzling. Few lines were visible, and those that could be seen were emission (bright) lines rather than absorption (dark) lines. Stars do not normally exhibit emission lines; emission lines are seen only in hot gaseous objects such as high temperature nebulae. Was the object surrounded by a hot gaseous envelope? Furthermore, Sandage couldn't identify the lines; they didn't make any sense. To him, the object was an enigma; it looked like a star with a strange spectrum, perhaps surrounded by a cloud of hot gas.

Jesse Greenstein of Caltech was soon drawn into the mystery. After obtaining his Ph.D. from Harvard in 1937, Greenstein worked for a few years at Yerkes Observatory. From there he went to Caltech, and in 1965 he was appointed chairman of the Division of Physics, Mathematics, and Astronomy. His major interest was the constitution of stars.

With a strong background in physics and spectroscopy, considerable experience in dealing with the effects of temperature and pressure on spectral lines, and a strong determination, Greenstein was the right man for the job. He soon became convinced that 3C 48 was an odd star—perhaps a supernova remnant, but different from most. He worked for months on the spectrum, puzzling it out, sketching models of atoms that might produce such lines. He became convinced that the lines were due to heavy elements not usually seen in stars. But he wasn't sure. Something kept nagging at him. Was he missing something, he asked himself? Somehow he felt that he had. Reluctantly he wrote up a lengthy research report for the *Astrophysical Journal*. Before sending it in, however, he gave it to a fellow astronomer, Maarten Schmidt, to check over. Schmidt could see nothing wrong with it, so Greenstein put it in the mail.

INSIGHT

Meanwhile at Cambridge University a young radio astronomer, Cyril Hazard, had also become interested in identifying radio sources in the 3C catalog. He searched for a method to identify them optically but, with the poor resolution of radio telescopes, it seemed impossible. Then he noticed that one of the most intense sources, 3C 273, was going to be occulted (eclipsed) by the moon. In fact, it would be occulted three times within a few months. When the occultation took place there should be a sharp cutoff of the radio waves, he reasoned. This would allow him to narrow in on its position.

Shortly thereafter Hazard moved to Parkes, Australia, where he had access to the newly completed 210-foot Parkes

Cyril Hazard.

dish. He was eager to try out his new idea. But as the date of the first occultation approached he made a disappointing discovery. It was out of range of the dish—but just barely out of range. Talking it over with the director of the observatory, John Bolton, they decided that a trench could be dug beneath the dish to allow it to access positions near the horizon where the occultation would take place. And, luckily, the technique worked, and the coordinates of 3C 273 were obtained.

Hazard sent the coordinates to Martin Schmidt at Palomar, asking him to try to make an optical identification. Schmidt had only been at Palomar a short time when he received Hazard's letter. He had first visited Palomar after obtaining his Ph.D. from the University of Leiden in 1956. But there had been no permanent position available at the time so he returned to the Netherlands after about two years. He had barely returned, however, when Minkowski retired and a position became available.

Schmidt's interest in astronomy started during World War II when he was barely in his teens. Night after night there were blackouts as bombers from England roared overhead. Looking up at them Schmidt also saw bright stars, and he soon became intrigued with them. Encouraged by his grandfather, he began reading about astronomy, and eventually made a crude telescope. Between bombing raids he used it to study the stars and planets.

Schmidt was heavily involved with other projects when the coordinates arrived from Hazard, so he had to make time between other projects. Looking up the coordinates on some of the available plates he and Matthews were able to find an object. As in the case of 3C 48, it looked like a dim bluish star, only this time it was a little brighter. Furthermore, there was another difference: to one side of the star was a small jet that looked as if it might have been blown out of it. Was the object an exploding star—perhaps a supernova remnant? It didn't look like the usual supernova.

It was an intriguing object, but he couldn't drop everything to study it. After all, it could turn out to be nothing more than a simple radio star. In late December, 1962, after he had completed his previously scheduled work on the 200-inch telescope, he decided to take its spectrum. But what was the important part: the star or the jet? He decided to take a spectrum of both.

The next morning, after developing the spectrum of the star, he held it up to the light. He had overexposed it. Then he looked at the spectrum of the jet. The plate was blank. A waste of several hours. A few weeks later he decided to try again. This time he shortened the exposure and, indeed, when he developed the plate, several lines were visible. Like 3C 48 they too

Martin Schmidt.

were strange: They were emission lines and didn't appear to correspond to any known lines.

Schmidt showed the spectrum to Greenstein, Sandage, and Matthews, but none of them knew what to make of it. More lines were visible in this spectrum than there were in the spectrum of 3C 48. But still, they made little sense. Schmidt puzzled over them but couldn't figure out where they came from. In February, *Nature* magazine in England wrote him asking him to submit a short report on the strange object. He agreed, but was uncertain what he was going to say. What was there to say?

Earlier, he and Greenstein had speculated that the lines had undergone a large Doppler shift toward the red end of the spectrum (a redshift). But they had laughed at such outlandish speculation. It didn't seem possible that a star could have such a huge redshift. Only galaxies had large redshifts. Yet, as Schmidt studied the lines he realized that they looked vaguely familiar. In fact, with the exception of a couple of the lines, they looked like a series of lines that appear in the spectrum of hydrogen, called the Balmer series.

Schmidt took out his slide rule and did some calculations. The separation of the lines matched the Balmer series separation. But what were the other two lines? Thumbing through a table of spectroscopic lines, he found what he was looking for: They could be magnesium and oxygen lines that had been redshifted. Things were finally beginning to make sense. A large redshift could also explain why the Balmer series was in the wrong position. Furthermore, the same redshift made sense of the other lines. His slide rule told him the redshift would have to be 16%—unheard of for stars. But, after all, this was a peculiar star—if a star at all.

A redshift of 16% meant that the object was traveling at about 15% the speed of light and, if Hubble's interpretation for galaxies was used, it placed the object roughly 2 billion light-years away from us. It couldn't be a star. It had to be some sort of extragalactic object.

At that moment Jesse Greenstein passed by the open door of Schmidt's Caltech office. "Jesse," he called to him. "Come in here." He told him what he had concluded. Greenstein thought about it for a few moments, then suddenly a strange look came over his face. "If that's the case," he said. "3C 48 has a redshift of 37%." With that they both rushed back to Greenstein's office to look at the spectrum of 3C 48. And sure enough, he was right.

It took a while for the results to sink in. Schmidt and Greenstein made further checks. Soon they were joined by Matthews.

These were not stars, they were a new type of object. And it was possible that there were many more of them—some with even larger redshifts. Schmidt quickly wrote up a paper for publication. It was a short two-page paper but it is now considered to be a classic: the announcement of a new type of object. Greenstein quickly withdrew his paper on 3C 48 and rewrote it.

THE RACE WAS ON

Astronomers had no idea what the objects were. They looked like stars, yet were as distant as some of the most distant galaxies. The began calling them quasi-stellars, indicating that they were like stars, but different. This was soon changed to quasars.

The race was soon on to find more of these objects, and more were soon found. In 1964 Sandage and Ryle developed a technique that allowed them to identify the objects relatively quickly. They noticed that quasars emitted a large amount of ultraviolet (UV) radiation (most stars emit very little UV). Sandage and Ryle therefore photographed sections of the sky using a UV filter, then they shifted the plate slightly and photographed the same region using a blue filter. This gave them a series of double images that allowed them to identify quasar candidates. (They qualified if they had UV images brighter than their blue images.) The final determination was done by studying their spectrum.

Through this method Sandage and Ryle found many new quasars. To their surprise, in fact, they found many objects with excess UV that were not radio sources. And, indeed, we now know that most quasars are not radio sources. Only about 10% are. All, however, are strong emitters in other regions of the electromagnetic spectrum, such as X ray, infrared, and so on.

With this new technique astronomers began finding quasars with larger and larger redshifts. The real shock, however, came in 1965 when Schmidt announced that he had discovered a quasar (3C 9) with a redshift that indicated it was speeding away from the Earth at 80% the speed of light. Accord-

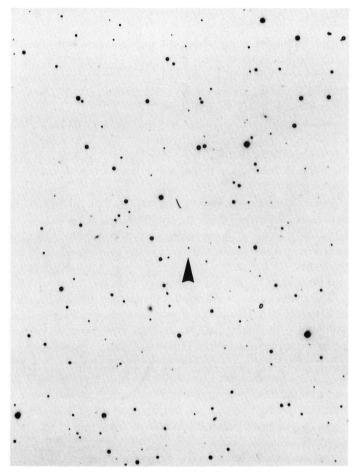

Quasar (above arrow). (Hale Observatories)

ing to the Hubble plot, the object was 10.5 billion light-years away. This was over double that of 3C 48. But not only did it double the size of the universe, it doubled its age. Astronomers were astounded. And they still had little idea what the objects were.

ENERGY

The most amazing thing about the quasars, however, was their energy output. If they were, indeed, as far out as their redshift indicated (according to the Hubble plot) they were giving off a hundred times as much energy as a galaxy. Yet a galaxy typically consisted of 100 billion stars.

Incidentally, when I refer to their distance as indicated by the Hubble plot I'm assuming that they are "cosmological." In other words, like galaxies, they were thrown outward by the big bang explosion, and are therefore taking part in the overall expansion of the universe.

But if quasars are indeed as far away as 10 billion light-years, how could we be seeing them? A galaxy, which is made up typically of 100 billion stars, looks like a point at this distance. We can distinguish it only because of its slightly fuzzy appearance. Yet quasars look like stars. There is no way we could be seeing a single star at that distance.

The problem was compounded when, shortly thereafter, it was discovered that some of the quasars were variable. A large object such as a galaxy cannot vary its light over a short period, say a month, or even a year. The reason is that the various parts of the galaxy have to be synchronized to cause such a variation. If a galaxy, for example, is 100,000 light-years across, it can't change its light output significantly in under 100,000 years. The reason: according to relativity, nothing can travel faster than the speed of light, and it takes 100,000 years for a light beam to cross the galaxy.

But quasars were changing their brightness significantly in months, even weeks and days. This meant that they had to be small. If an object is able to, say, double its light output in a day, it can be no larger than a light-day across. And some quasars were changing significantly in a day. This meant that they could be no larger than our solar system.

How could an object this small give off more energy than an entire galaxy? One way around the problem was to assume that quasars were not cosmological. It was possible that they could have speeds close to that of light and not be extremely distant. If this was the case, however, we would have to explain where their high speed came from. James Terrell of Los Alamos Scientific Laboratory suggested that they might have been blown out of our galaxy at high speed. There are indications that some sort of explosion is taking place near the core of our galaxy. But if this was the case, some of the quasars should be moving toward us, and their spectral lines would be shifted toward the blue end of the spectrum. But we observe no blueshifts; all quasars exhibit redshifts.

To overcome this objection, the suggestion was made that the explosion in our galaxy occurred a long time ago, and all quasars are now in the outer regions of our galaxy—out past us. But it was soon discovered that such an explosion would deplete a galaxy of almost all its energy. And there was no indication that our galaxy was energy-depleted. Besides, nearby galaxies should also be evicting quasars, and some of them would be coming our way.

Halton Arp, who was at Palomar at the time, uncovered another problem with the idea. He exhibited a photograph of a galaxy and a quasar with significantly different redshifts that appeared to have a bridge between them. Is this possible? Obviously it isn't. Assuming they are cosmological, all objects with the same redshift are at the same distance from us.

Even if quasars are nearby—and few astronomers now believe this is the case—they are still extremely energetic. Where is all this energy coming from? In the last chapter we saw that radio galaxies are also energetic and we hypothesized that they might have a gigantic black hole in their nucleus. Is this also the case for quasars? Most astronomers believe it is.

THE BLACK HOLE MODEL

The energy output from quasars is tremendous and, as mentioned earlier, although most quasars are not radio sources, all are strong sources of X rays, UV, infrared, and so on. Further-

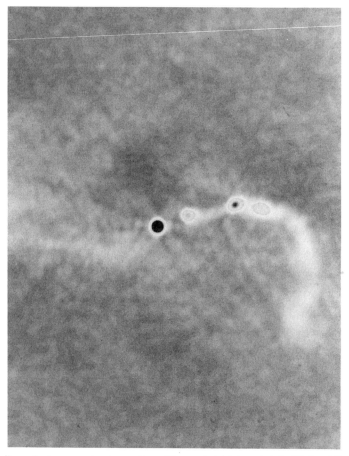

Radiograph of quasar showing radio-emitting material being ejected. [National Radio Astronomy Observatory (NRAO) and observers J. J. Condon and H. S. Murdoch]

more, they are exceedingly small. One of the only small sources that can supply such amounts of energy is a black hole. As in the case of radio galaxies, this black hole would have to be extremely massive—several billion times as massive as the sun.

But a black hole all by itself is not an energy source. It needs fuel. Would fuel be available? It is likely that large numbers of stars and a considerable amount of gas orbit the quasar. They could fuel it. In fact, it is quite possible that the stars and gas would be in a huge accretion ring. If we assume this is the case, viscous friction in the material of the ring would cause it to move closer and closer to the black hole, until finally it plunged into it. In the final seconds before the plunge, as it was stretched and heated by the tidal forces, it would give off X rays and other radiations. Furthermore, huge jets might be produced, just as they are in the case of radio galaxies. In fact, 3C 273 has a visible jet.

I asked Richard Green of the National Optical Astronomical Observatory (NOAO) if he thought it likely that most quasars have jets. "It's assumed that many do," he said. "The question is: How far out from the system do they get? When you look at powerful radio quasars you see little jets all the way down to the smallest scales . . . so the phenomenon looks fairly common. When the quasar isn't a radio source, though, it's a more open question as to whether there is a jet. Maybe there is but it doesn't get far before it is absorbed by the material of the quasar."

While the black hole model of the energy of quasars is the best we have, it's not problem-free. Daniel Weedman of Pennsylvania State University pointed out several problems. Weedman was an undergraduate at Vanderbilt University, and a graduate student at the University of Wisconsin where he worked under Donald Osterbrook on planetary nebulae. He first became interested in quasars while he was at the University of Texas studying a group of active galaxies called Markarian galaxies. Soviet astronomer E. Khachikian and Weedman began observing them. "A lot of the Markarians turned out to be Seyfert galaxies," said Weedman. "This eventually led me to work on quasars."

"The black hole model is really the only model we've got. Nothing else comes close," said Weedman. "One of the problems with it is that, as time passes, black holes should grow

Daniel Weedman.

more massive, and you would therefore think that quasars would become more powerful as they got older, but in fact just the opposite is true. The most powerful quasars are the very bright young quasars."

He went on to talk about the problem of the X-ray spectra. The highest-energy X rays should come from close to the black hole and their variation should give an indication of the size of the black hole. Rapid variations have been seen, he said, but they do not indicate a size appropriate to gigantic black holes.

"The major problem with the black hole model, however," said Weedman, "is that you can never directly confirm it. The only indication you have of it is the gravity it exerts, and this occurs on such an extremely small scale. We simply can't observe quasars in enough detail to prove unambiguously that they have a black hole at their center."

Richard Green also has slight reservations with the black hole model. "Gravitational energy release is certainly a very attractive candidate for the kind of power that is required," he said. "And as long as you don't look in excruciating detail as to how the energy of gravitation is transformed to the radiant energy we see, the model looks like it does a reasonably good job of explaining things. But I think the details of the model are far from proven. The details of the energy production have encountered a lot of practical problems, but I don't think they invalidate the model."

QUASARS WITH FUZZ

In 1979 Susan Wyckoff of Arizona State University decided to take a close look at 3C 273 using the latest equipment. In particular, she wanted to use computer processing of the image to see the details of its structure. Working with her husband, Peter Wehinger, she traveled to the European Southern Observatory in Chile to get the required data.

"The data was computer-processed to bring out the low-surface brightness regions and the quasar image," Wyckoff said. "It was in the computer processing that we found evidence for a galaxy . . . or at least 'stuff' around the quasar. And the characteristics such as surface brightness and so on, of the stuff turned out to be what we would expect galaxies to have at the distances indicated by the redshift of the quasar. So we concluded that the quasars were simply very bright nuclei of distant galaxies." The reason we don't normally see the surrounding galaxy, she said, is that the quasar nucleus is so bright it swamps the light from the galaxy, making it difficult to see.

I asked her if she is convinced that all quasars are bright nuclei of galaxies. "I would say that's the general consensus." she replied. "However, that's not to say that there isn't another kind of quasar in the universe . . . an entirely different object. We haven't examined every quasar in that much detail so we don't know if there is a second population of them."

Susan Wyckoff.

Most astronomers are now convinced that quasars are, indeed, the nuclei of distant galaxies. "That doesn't mean we understand how they got there," said Weedman, "but at least we know that they ought to be identified with galaxies."

Richard Green concurs. He feels that most quasars are associated with galaxies but has slight reservations about the most distant ones. "Whether the ones in the very early universe—the very highest redshift ones—have developed a full-fledged galaxy around them is still an open question. But I have the feeling that the kind of gravity that is required almost guarantees it's going to be a galaxy-like object."

EVOLUTION

One of the major problems in relation to quasars is how they evolve, in other words, how they change as they age. There's no question that they do change, since we see the most distant ones at an earlier stage of their life than those closer because of the finite speed of light. And these young ones are, in general, much more powerful.

Closely related to the problem of evolution is the relationship between the various types of galaxies: quasars, Seyfert galaxies, and radio galaxies. What is their connection? Weedman is confident that quasars and Seyferts are the same thing. "Seyfert galaxies are just so close that we see the galaxy part in addition to the quasar in the nucleus," he said. "The only puzzle is whether or not all galaxies have gigantic black holes at their center. Probably they do. But whether or not quasars are related to all galaxies as opposed to just Seyfert galaxies, depends on whether galaxies in general have black holes at their center."

Richard Green, along with Howard Yee of the University of Montreal, has done a considerable amount of work on the evolution of quasars. Green did his doctoral thesis under Maarten Schmidt of Caltech. "Originally, I was going to do my thesis on quasar evolution," he said, "but after surveying about a quarter of the sky looking for quasars, I found a lot more white dwarfs on the plates than I did quasars, so I did my thesis on the distribution of white dwarfs in the neighborhood of the sun." After his thesis was complete, however, he stayed on as a research assistant and eventually worked on quasars. Since then he has done a considerable amount of work on the evolution of quasars. His most recent work with Yee has been a study of quasars that are associated with clusters of galaxies. They have been trying to understand what it is that creates quasars in one place (say, a cluster) and not another, and how this is related to their evolution.

"What Howard Yee and I found in our research," said Green, "is that quasars that are found at low redshifts, corre-

Richard Green.

sponding to times less than one-third the age of the universe [back in time from the present], are usually found in groups of galaxies. In fact, as strange as it seems, they are three times more likely to be associated with a galaxy than another galaxy is." In an effort to clarify, he gave me an example. "If you look at a galaxy in the sky, there is some probability that it's got a neighboring galaxy. A quasar is three times more likely to have a neighboring galaxy." He paused. "Furthermore, radio quasars are found in nearby dense groups, but not in the cores of giant clusters. But if you look at these same powerful radio quasars at greater distances [and earlier times], you find a substantial fraction, approximately one-half, in the cores of giant clusters of galaxies—not just small groups. So there's something about the centers of these clusters that was different at early times.

They're still there at later times, but they no longer contain powerful quasars."

Green then went on to describe some of the work on the same project that a graduate student, Erica Ellingson, who is working for him, has been doing. "She looked at the critical redshift region where we saw the transition. She found that, as you came to lower redshifts, the quasars that were in the clusters were fainter and fainter. She found a handful in the giant clusters, but they were extremely faint."

Green and Yee have proposed two different ideas to explain these results. Their first proposal is that quasar activity may be triggered when two galaxies collide in a special way. "The best model is a galaxy passing by a quasar that rearranges some of the gas clouds surrounding it so that they fall into the middle and create fuel for the quasar. For this kind of collision to be effective, though, the two objects have to move fairly slow." He paused. "What's happening in the giant clusters, we suspect, is that they are becoming denser in the center with time. This means the galaxies near the center have to speed up and they therefore pass one another quickly. This kind of collision is not likely to be effective in fueling the quasar. That's why quasars are not found in giant clusters at late times."

Green and Yee's second proposal assumes that there is a fuel source within the host galaxy that fuels the quasar. "The very rich clusters are often powerful X-ray sources," said Green. "This means that individual galaxies in the middle of clusters do not hold onto very much gas. The gas is somehow scoured out. So there's not enough fuel in the center of the galaxy at later times to light up the quasar."

DO COLLISIONS TRIGGER QUASARS?

As we saw in the last section one of Green's ideas on the evolution of quasars is that collisions between galaxies trigger quasar activity. In the last few years this has become of considerable

interest. Besides Green and Yee, two other groups have supplied strong evidence to support this point of view. Both of
these groups feel that collisions fuel a massive black hole, triggering a quasar to life.

Patricia Vader, Gay Da Costa, and Charlene Heisler of Yale
University, Michal Simon of the State University of New York at
Stony Brook, and Jay Frogel of NOAO have studied a relatively
nearby quasar (approximately 4 billion light-years away) that
appears to be colliding with a galaxy. It was detected by the
Infrared Astronomical Satellite and is called IRAS 00275-2859. It
appears to be a collision in its early stages, but there is evidence
that the shock of the collision has set off a burst of star formation
in the quasar's host galaxy. The quasar and the nucleus of the
galaxy are quite close to one another (projected distance is approximately 36,000 light-years), yet there is little deformation.
There are indications, however, that the quasar has been formed
by the collision. Gas from the galaxy, it is believed, has found its
way to the nucleus of the quasar's host galaxy, triggering the
quasar and making it luminous.

The other case, referred to as Markarian 231, is being studied by Donald Hamilton and William Keel of Palomar and the
University of Alabama, respectively. This collision is at a much
later stage. It is also closer, lying less than a billion light-years
away. The quasar is powerful and is surrounded by a dense dust
cloud. A blast wave appears to be emanating from the nucleus.
Furthermore, the galaxy is peculiar—it has no definite structure
and is quite lumpy.

I asked Weedman if he is convinced that collisions of this
type could trigger, or create, quasars. "It's easy to understand
how it could happen early on," he said. "Anything that can
apply a new source of fuel to the black hole would cause the
quasar to flare up. When two galaxies collide, the gas in the
galaxies is disrupted in such a way that some of it would get into
the vicinity of the black hole. So it's fairly easy to see how galaxy
collisions could trigger quasars. But that doesn't explain how
the black hole got there in the first place."

SPECTRA

The spectral lines of quasars were an enigma when they were first discovered, and in some ways they are still an enigma today. The first quasars all exhibited emission lines, but in 1966 a quasar was discovered that also had absorption lines. The problem, however, was not that there were two types of lines in the same spectrum. What was puzzling was that the two different types of lines had different redshifts, indicating that they were at different distances. Furthermore, when the absorption lines were studied in more detail astronomers found that they exhibited more than one redshift. Some quasars, in fact, had as many as six.

What was causing this? Astronomers now feel that they have a reasonable explanation. First of all, they believe that all of the emission lines are produced by the quasar itself, but that some of the absorption lines are not. Some of the absorption lines are likely due to galaxies and gas in the line of sight between us and the quasar. But what about the large number of different redshifts? To explain them they have had to assume that some of the absorption lines are due to expanding clouds of gas in the galaxy around the quasar, at different distances from it. There may, in fact, be several cloud layers—each expanding at a different rate.

"The problem with the absorption lines," said Weedman, "is that you get some from the quasar and some from the intervening material. It's difficult to sort them out. There will always be controversy as to whether any particular absorption line is intervening, or belongs to the quasar. The most interesting aspect of the interstellar absorption lines is that they are associated with galaxies and other things that we cannot see in any other way. In that respect quasars are very important in providing background searchlights that shine through galaxies and other clouds of material along our line of sight. I think most of the future observational research related to absorption lines will be directed towards these intervening lines."

Wyckoff is now working on this problem. "We're trying to image and identify the intervening objects," she said. "Even though we don't pick the objects up in deep sky images we see strong evidence for them in the spectra of quasars."

THE CUTOFF

When Schmidt discovered a quasar that was receding from us at 80 percent the speed of light in 1963, astronomers were shocked. Soon there was a rush to find even more distant quasars. But how, you might ask, do astronomers label distances? It turns out that they use what is called a z number. If the redshift displaces the spectral line by a certain amount (equal to the magnitude of the wavelength), the quasar has $z=1$. If it is displaced by twice this amount, it has $z=2$, and so on. Schmidt's 1963 quasar had a z of 2. And throughout the 1960s astronomers found many quasars with redshifts greater than 2. But no one could find one with $z=3$. Finally, however, in the early 1970s the first quasar with z greater than 3 was found. It had $z=3.4$, but later one was found with $z=3.53$, corresponding to a distance of 11.5 billion light-years. For 10 years this quasar was the most distant one known.

This leads us to ask: Is there an eventual cutoff distance for quasars? If you think about it for a moment, it's easy to see that there should be. If we assume that the universe began with the big bang, it was originally composed of a gas cloud of particles. At that time there were no galaxies, or quasars, in the universe. They hadn't been formed yet.

Strangely, though, we seem to keep pushing this cutoff back farther and farther. Although no quasars with $z=4$ were found for 10 years, even this barrier was broken in 1987. In fact, a total of 7 quasars with z greater than 4 were discovered. The most distant of these was one that was found in September 1987, by Stephen Warren and Paul Hewitt of Cambridge University. It has $z=4.43$ and is at a distance of 12 billion light-years.

The most distant quasar known at the present time is one that was discovered in 1989 by Donald Schneider and several colleagues at the Institute for Advanced Study. It has a z of 4.73.

The breakthrough to z greater than 4 was primarily due to the development of a new optical technique for identifying distant quasars. In this method, called the Automatic Plate Measuring Facility (APM), an instrument reads the position and magnitude of up to 200,000 stars and galaxies a day. Plates are taken with several different color filters and the APM compares them and identifies potential quasars.

Interestingly, though, all the quasars with z greater than 4 were not discovered using the APM. On September 25, 1987, Patrick McCarty and Mark Dickinson were trying to obtain the spectrum of a radio galaxy at Lick Observatory. In fact, they wanted two spectra—one of the radio galaxy, and one of a nearby companion. They therefore placed the slit of the spectrograph so it would cover both objects. But during the exposure, the other end of the slit caught the spectrum of a quasar. Measuring the redshift of the quasar, they were amazed. It was 4.40.

Richard Green reminded me, however, that even with the z=4 barrier being broken, there is still a tremendous cutoff beyond about z=2.5. "The number of quasars you find per unit volume falls off exponentially as you look back in time. There is a peak of quasars at about 80 percent the age of the universe back in time. At times farther back there are only a handful . . . physicists would certainly call this a cutoff."

THE FUTURE

A tremendous boost for quasar astronomy came when the space telescope was put into orbit in 1990. Weedman, who is involved with the project, described some of the projects that are planned. "The absorption line problem is one of the first things that will be done. We will try to find intervening dark galaxies that are quite close to us which are seen only through their absorption lines. We will also use quasars to study the gravitational lens. In other words, we will try to find quasars that have double images because a lot of their light is being lensed by dark matter. And the third, extremely important project, will be to look at the centers of nearby galaxies—galaxies that are not currently known to be quasars to see if there is any evidence for tiny quasars in their centers, and also to see if there is any evidence for black holes at their centers."

Green is also looking forward to the space telescope. "With the sharper resolution, we should be able to see the center regions of Seyferts clearly," he said. "Just getting the sizes and geometries of the centers of some of them will be a tremendous help."

The space telescope will also give us better access to the UV, and UV light is an important component of quasar light. "Nature has blessed us with a lot of atoms that give up most of their information in the UV so we need that kind of spectroscopy," said Green.

CHAPTER 7

Is Our Galaxy Exploding?

We've seen evidence for explosive power in the centers of quasars, Seyferts, and other radio galaxies. Jets and other evidence of an explosion are actually visible in some cases. While galaxies such as ours show little evidence of similar power, our galaxy does appear to have something mysterious going on at its core. Its core is by far its most energetic region, and in many ways, its most bizarre. Jansky and, later, Reber, found it to be a strong source of radio waves, but we now know that it is also a strong source of infrared radiation, ultraviolet radiation, and X rays.

But if it is emitting so much energy, what is causing it? How is it generated? Is it possible that there is a black hole here as there appear to be in the cores of quasars and other radio galaxies?

STRUCTURE

By the early 1920s a considerable amount had been learned about the structure of our galaxy. But there was still a lot of controversy. Two models were at center stage. They were in agreement that our galaxy was a large disklike structure, but they disagreed as to its size. Furthermore, they disagreed as to where the sun was located.

The first of these models was constructed by the Dutch

astronomer Jacobus Kapteyn. In the early 1900s he organized a worldwide effort aimed at determining the structure of our galaxy. Astronomers made observations in both the northern and southern hemispheres. Kapteyn published his results in 1922. According to them our galaxy was approximately 30,000 light-years across, and about 6000 light-years thick. And our sun was located at its center. The size was not a surprise, for a similar estimate had been made in the late 1700s by William Herschel.

But not everyone agreed with Kapteyn's model. Harlow Shapley of Mount Wilson Observatory had been using globular clusters to study our galaxy since 1916. (They are clusters of stars that contain from a few hundred thousand to a few million members.) He determined that our galaxy was much larger than Kapteyn believed it was—ten times larger, in fact. Furthermore, his observations indicated that the sun wasn't at the center, but out near the edge. Some astronomers were outraged by Shapley's model. It wasn't his claim that we were at the edge that bothered them. But the size he got, they were sure, was beyond reason.

Shapley discussed his model with Kapteyn, showing him some of his results. But Kapteyn was not convinced. He was sure an error had been made. And Shapley was equally sure he hadn't made one.

Kapteyn worked almost to the day he died on his model. Near the end of his life he asked one of his graduate students, Jan Oort, to look into the motions of stars in the Milky Way. Astronomers had always assumed they were random, but Kapteyn noticed during his study that there was some regularity. Most were moving in the direction of the constellation Orion, but some were moving in the opposite direction. He wanted Oort to study the fastest moving stars as part of his doctoral thesis.

Oort found it to be a difficult and frustrating problem. A number of the stars were, indeed, moving much faster than the others, and the fastest ones were all heading in the same direction. Furthermore, these results were at odds with Kapteyn's model. Oort finally decided that the fastest moving ones were

Bertil Lindblad.

just "interlopers"—intruders with highly inclined orbits. But he wasn't satisfied with this explanation.

The Swedish astronomer Bertil Lindblad read about Oort's results. He had completed his Ph.D. at Uppsala only a few years earlier. He had also traveled through the United States, working briefly at Mount Wilson, Lick, and Harvard observatories. So he knew all about the controversy between the Shapley and Kapteyn models.

Lindblad, however, was not an observer. He was a theorist with a serious interest in the dynamics of the Milky Way galaxy. How did it rotate? And why? Applying mathematical methods to the problem he found that Oort's interlopers were not high-speed stars. They only appeared to be moving fast because the sun was overtaking and passing them. There was, in fact, nothing mysterious about them at all.

When Lindblad published his results in 1925 they soon

came to the attention of Oort. And he saw something that wasn't fully realized by Lindblad. The stars in the neighborhood of the sun were apparently moving like the planets of the solar system. In the solar system the inner planets move faster than the outer ones. In the same way, stars closer to the center of our galaxy than the sun seemed to move faster than the sun; those farther out moved slower. (We will see later, though, that this is not true throughout the galaxy. Its motion is quite complex.) Oort also showed that all the stars were rotating around a point in the direction of the constellation Sagittarius. Amazingly, it was the same point Shapley had shown years earlier to be the center of the globular clusters.

Lindblad continued to struggle with the Milky Way's motion for many years. He wanted to determine the structure of the spiral arms, and spent years making detailed calculations of the orbits of stars within the galaxy. Many of his results, however, have been forgotten, mainly because not enough observations were available at the time. In several cases, in fact, he came to wrong conclusions. He concluded, for example, that the arms were unwinding, rather than winding. The galaxies, he was sure, were not rotating with trailing arms as they appeared to be. He believed that they were rotating in the opposite direction, and therefore their arms were unwinding. Despite problems such as this, his contributions were important. One of them, in fact, was critical to a later understanding of the galaxy's form. He determined that it took much longer for the spiral arm to move around a galaxy than it did the stars. This led him to conclude that the arms were caused by a "density wave" that moved around the galaxy. A density wave is similar to a sound wave. When we talk, molecules of air move back and forth along the direction that the sound is traveling. The density wave within our galaxy would cause stars to move in the same way, according to Lindblad.

Meanwhile, the controversy between the Shapley and Kapteyn models continued. Kapteyn admitted, though, that he was not entirely satified with his model. He realized that it was

Robert Trumpler.

based on the assumption that our galaxy had no intervening matter between the stars. If there was an obscuring medium of any type his model was in trouble. He was also uneasy about the sun's position in it. Why would it be exactly in the center? The problems worried him.

His concern, it turned out, was justified. Within a short time an interstellar medium was discovered and Kapteyn's model was discarded. The astronomer who made the discovery was Robert Trumpler of Lick Observatory. Educated in Switzerland and Germany, Trumpler came to the United States in 1915, and soon joined the staff of Lick Observatory. His major interest was the structure of the Milky Way galaxy. He felt that one way of determining it was by studying the galactic clusters within it.

Trumpler therefore set out, using several techniques, to determine the distance to as many galactic clusters as possible. This would give him an indication of the Milky Way's size. In 1929 he published his results: It was about 35,000 light-years across—in fairly good agreement with Kapteyn's model. His next task was to determine the size of the galactic clusters themselves. Using the data he had obtained, he went to work. But before long he encountered something strange: His results indicated that the farther away a cluster, the larger it was. This made no sense. In fact, it was crazy. He checked his calculations but everything seemed to be correct. He thought about the result, trying to puzzle it out, but the only thing he could come up with, assuming the clusters did not increase in size, was that there was an obscuring medium throughout the disk of the galaxy. Taking the dimming that would occur into consideration, assuming this was the case, he found that the sizes stayed the same. But an obscuring medium would have disastrous effects on Kapteyn's model, and he was reluctant to give up the model. Nevertheless, in 1930 he published his results.

The importance of Trumpler's discovery was soon realized. Kapteyn's model was, indeed, incorrect. The sun is not at the center, but out near the edge, as Shapley had indicated. But Shapley had also assumed that there was no interstellar medium, so his model was also incorrect. Taking the medium into effect it was soon shown that Shapley's model was reduced in size by a factor of three.

Astronomers had now determined that our system was disk-shaped, with spiral arms, and it had a diameter of about 100,000 light-years. Also, it rotated like a gigantic wheel in space. But still there was much to be learned. Astronomers had no idea what the central region looked like, nor did they know the detailed structure of the arms. Answers to both of these questions were soon to come, however. The first steps came during World War II.

Walter Baade arrived at Mount Wilson from Germany just

before World War II. When the war broke out he was still an alien, so while most of the other astronomers went off to war-related projects, he was restricted to the observatory and the area around Pasadena. Oddly enough the restriction was a blessing in disguise. Baade had the 100-inch reflector almost to himself. Furthermore, the sky was free of the polluting glare from the lights of Los Angeles; they were blacked out because of the fear of an invasion.

Many years earlier Hubble had resolved the arms of the Andromeda and other galaxies. Stars stood out clearly on his photographs, yet the centers of the galaxies had never been resolved. Astronomers were particularly interested in resolving the center of the Andromeda galaxy—but they had not succeeded. Baade was convinced that conditions were now ideal; he was sure he could succeed where they had failed. He began by trying several long exposures with the usual blue-sensitive plates. But nothing showed up—only the bright continuum that others had obtained. A new, particularly sensitive red film had just been invented that allowed astronomers to make exceedingly long exposures. On the chance that some of the stars in this region were red, Baade decided to try it. When the first plates were developed he was astonished: stars stood out like tiny rubies. He had resolved the core! But, more importantly, he had used red-sensitive plates to do it. This meant that the stars in the region were red, and therefore likely old.

But most of the stars in the arms, at least the ones that stood out, were blue stars. Baade concluded therefore that there were two kinds, or populations, of stars in the Andromeda galaxy. He referred to them as Population I (abbreviated Pop I) and Population II. The young blue stars in the arms were Pop I and the older red stars in the core were Pop II.

Did this also apply to the Milky Way? Baade later showed that it did. We now know, in fact, that all spirals are made up of Pop I in the arms and Pop II in the core (the globular clusters that surround the galaxy are also Pop II).

William Morgan of Yerkes Observatory in Wisconsin real-

Part of the core of the Andromeda galaxy showing Pop II stars. (Lick Observatory, University of California, Santa Cruz, Calif. 95064)

ized that Baade's discovery was an important breakthrough. Using it he would be able to map the arms of our galaxy. Along with several students he therefore began determining the distance to blue giants in the arms. And by 1951 he was able to make a crude map of some of the arms.

A spiral galaxy similar to the Milky Way galaxy. (Hale Observatories)

THE RADIO VIEW

The first real progress in mapping the arms of the Milky Way galaxy, however, came from radio astronomers. Reber's map of the radio waves from the core of our galaxy was published during World War II. Although Holland was under Nazi occupation at the time, Oort managed to get a copy of it. He could not make observations because of the occupation; nevertheless, he saw that radio astronomy had tremendous potential. He felt that it might be of considerable help in determining the structure of our galaxy. The key, he was sure, was hydrogen. Our galaxy is full of hydrogen, and if it gives off radio waves we should be able to detect them. He asked a colleague, Hendrik van de Hulst, to check and see if hydrogen gave off radiation that could be detected. Van de Hulst decided to tackle the problem using quantum theory. After a number of blind alleys it came to him: hydrogen atoms spin, and if they changed the direction of their spin they might give off radiation. He made the appropriate calculations and found that they would, indeed, radiate.

Let's look at how he came to his conclusion. As I just mentioned, hydrogen atoms spin. Assume for sake of illustration that they spin in a counterclockwise direction. Van de Hulst showed that occasionally they would make a "spin flip." In other words, they would suddenly start rotating in a clockwise direction. And when they made this spin flip they would give off radiation of wavelength 21 cm. They would only make such a flip once in about 10 million years, but because of the large number of hydrogen atoms in space, many would be flipping at any given time.

Van de Hulst and Oort were not able to check on the prediction during the war, but as soon as it was over they organized a group and started work. By now, though, the prediction was known in other parts of the world. Then disaster struck van de Hulst and Oort's group; they had barely gotten started when a fire destroyed much of their equipment. Meanwhile, in the United States, Edward Purcell and Harold Ewen made a small, crude antenna of plywood and copper foil. And in March 1951, they detected a radio wave of wavelength 21 cm. Oort soon heard of the discovery. Disappointed, he picked up the pieces, and quickly confirmed it. Confirmation also soon came from a group in Australia.

A valuable new tool was now available. But strangely, the Americans did not follow up on the discovery. This was left to

Oort and his group, and to various groups in Australia. The arms of the Milky Way were strewn with hydrogen, and the 21-cm radiation signaled its presence. Soon astronomers had detailed maps of the arms.

Let's look briefly at what they found. The Milky Way, it turns out, consists of three (or perhaps four—it's still controversial) arms wound one around the other. The arm between us and the center is known as the Sagittarius Arm (named for the constellation Sagittarius, which lies in this direction). Beyond us is the Perseus Arm, and between them is the Cygnus Arm. We are close to the Cygnus Arm, but are actually on a small spur that emanates from it, called the Orion Spur. There may be a fourth arm hidden from our view by the dense core of the galaxy.

What causes the arms? Even a casual glance at a galaxy such as the one in Andromeda appears to indicate that the arms are winding up. If the stars in the inner part of the galaxy move faster than those farther out we would expect a spiral pattern to arise. If, for example, you poured a black streak across the top of a can of white paint, then started stirring it at the center, you would soon have a spiral pattern. Is this the way the spiral arms of our galaxy arose? Let's look into this. We know that the sun's orbital velocity carries it around our galaxy once about every 250 million years. And since the sun is about 5 billion years old, it has made about 20 trips around our galaxy. But our galaxy is much older than our sun; it has probably made 60 or more full turns since it was formed. If this is the case the arms would not be relatively loosely wound, as they are. Imagine stirring the center of the white paint 60 times. Although a spiral pattern would emerge after a few turns, it would certainly disappear long before 60 turns were up. In 60 turns the arms of a galaxy would wind up so tightly we would not be able to distinguish it from an elliptical galaxy.

But if this is not the explanation, what is? The first clue had already been planted. Earlier I mentioned that Lindblad suggested there might be a density wave moving around our galaxy. In the 1960s the idea was resurrected by the American astronomers C. C. Lin and F. H. Shu. They assumed a density wave swept around our galaxy, but they focused their attention on the gas and dust in it rather than the stars, as Lindblad had. They concluded that it was the luminous gas that made the arms stand out. According to their calculations, as the density wave moves around our galaxy, gas and dust are swept into some areas, and out of others, and as the gas clumps, stars form—in particular, large blue stars. These large stars excite the gas around them and make the arms luminous.

The Lin–Shu theory was quite successful, but it didn't get around all of the problems. It didn't, for example, form stars fast enough. Modifications were obviously needed. And in 1966 an important one came. M. Fujimoto of Japan showed that the density wave was actually supersonic—it traveled faster than the speed of sound in the medium. This meant that there was a shock wave associated with the leading edge of the arm, and stars would now form faster.

But even with these modifications the theory could not explain all spirals. Many spirals are not perfectly regular: Some have fragments, others have feathered regions and spurs. Because of this, some astronomers are convinced that a few spirals may have formed differently. They have suggested that a chain reaction of exploding stars, or supernovae, starting near the center, could be responsible for their arms. Computer simulations, in fact, show that this is possible. A supernova near the core, for example, would cause a compression in the interstellar gas farther out. A large star would form here and within a relatively short time it would supernova, causing another compression beyond it. There would soon be a string of bright stars and clouds, and as the galaxy rotated this string would be sheared, giving it a spiral shape. Since massive blue stars are short-lived, the arm would fade before it got wound up extensively.

It is also possible that some of the arms were formed when galaxies passed close to one another. Powerful gravitational forces between them would pull out streams of gas and stars.

A dense star cloud in the direction of the center of our galaxy. (Lick Observatory, University of California, Santa Cruz, Calif. 95064)

Computer simulations of such collisions indicate that it would also create arms.

CLOSING IN ON THE CORE

So far I've said little about the core of our galaxy. We know that the center is in the direction of the constellation Sagittarius. If we look in this direction with optical telescopes we see a large number of stars, but in reality we are seeing only a tiny fraction of what is actually there. The core is cut off from our view by the gas and dust between us and it. Indeed, if this were not the case, this region of the sky would be exceedingly bright.

Fortunately we are not entirely cut off. Radio telescopes are able to penetrate the core to some degree. But even more valuable are infrared telescopes. They show a center a thousand times brighter than radio telescopes. Adequate intrumentation, however, didn't become available until the late 1960s, so little attention was paid to the core before that time. But once infrared telescopes were available astronomers made up for lost time. They soon found an incredibly energetic source at the center of our galaxy. As they studied it in further detail, they found that, in addition to radio and infrared waves, it also gave off X rays and gamma rays.

As a result of their efforts, astronomers now have a fairly good idea what the core looks like. Data from radiation of many different wavelengths have been correlated to give an optical model of the region. Many interesting features have been identified: huge clouds of gas, swirling rings of hydrogen, and perhaps even a black hole.

Like most spirals, our galaxy has a large nuclear bulge which extends out about 15,000 light-years from the center. Surrounding and enclosing it is a halo of huge molecular clouds. Further in, at about 10,000 light-years is a turbulent disk of hydrogen that is rotating at high speed. Strangely, this disk is tilted slightly (by about 20 degrees) to the overall plane of the

The region just south of the galactic center. (Lick Observatory, University of California, Santa Cruz, Calif. 95064)

galaxy. No one knows why, but we do know that it rotates at a speed of about 81 miles/sec. Also, it appears to be expanding outward, almost as if an explosion somewhere in the interior were pushing it out.

The disk, which consists mostly of molecular hydrogen, is

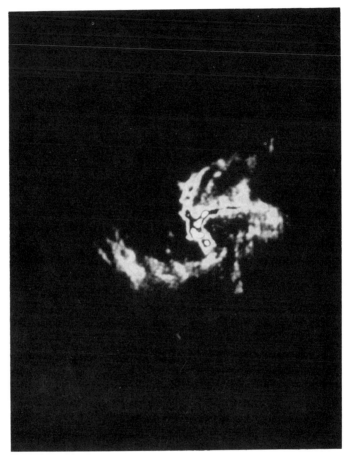

Radiograph of Sagittarius A. [National Radio Astronomy Observatory (NRAO) and observers K. Y. Lo and M. J. Claussen]

relatively cold in the outer parts. Closer to the center are a number of "smoke rings"—giant molecular clouds and gaseous nebulae that look like they have been blown out of the center. Closer yet are a number of radio sources, now referred to as

Tilted disk at core of our galaxy.

Sagittarius A, B, B2, and C (abbreviated Sgr A, B, B2, and C). Sgr A coincides with the center. More exactly, there is a small compact source called Sgr A* at its exact center.

Sgr A is an extremely intense source, giving off both thermal (heat) radiation and synchrotron (nonthermal) radiation. Sgr A* appears to emit only synchrotron radiation.

While radio astronomers have given us a lot of information about the core, it is infrared astronomy that has given us the real breakthrough. "The galactic center was saved for infrared astronomers," said George Rieke of the University of Arizona, who has been studying this region for many years. "The visible and ultraviolet astronomers have had free rein for years in the centers of other galaxies. But the one you can study the best was saved for us, because clouds of interstellar dust blot it out in the visible and the ultraviolet."

RED GIANTS NEAR THE CORE

Rieke grew up in Indiana and Illinois and went to graduate school at Harvard. "My father was a physicist and my mother an astronomer, so it was only natural that I became an astrophysicist," he said to me with a laugh. Using the 90-inch infrared telescope at Kitt Peak, he and his wife Marcia, who is also an astronomer, have discovered a number of red giant stars (large red stars of considerable brightness) very close to the core.

"There are some red giants within a few light years of the center," said Rieke. "But whether it's 3 or 4, or 5 or 6, is still not clear." He found evidence for them by studying the spectral lines of stars in the infrared.

I asked him how he got interested in the study. "There has been a controversy for a long time about what powers the galactic core," he said. "And I had noticed from earlier work that a number of sources at the center seemed to have small high-temperature cores suggesting that there were a number of internal heating sources rather than just one. And the most logical form of a number of heating sources was a collection of hot stars." He paused. "Of course, we can't see hot stars in the infrared. But if there are hot stars, stellar evolution tells us that there ought to be cool red supergiants to go along with them. So we set out to test this by taking the spectrum of the brightest stars we could see."

And, indeed, Rieke and his colleagues found evidence for red supergiants only a few light-years from the center. One star that they have studied in detail cannot be more than 10 million years old. This indicates that there has been a relatively recent burst of stellar formation in this region.

Astronomers have known for many years that the region within a few light-years of the center of our galaxy is a crowded place. "The real issue, though," said Rieke, "is whether the stars are old burned-out stars that have been here for the life of the galaxy, or whether there is a population of young stars." Rieke is convinced that they are young. Even though he has found only red supergiants, Rieke said that if you look at other regions where there has been recent star formation, you find roughly 10 blue stars for every red star. It seems reasonable, on the basis of this, that blue stars are present.

AT THE CORE

One of the most prominent features close to the core is a ring of whirling matter—mostly hydrogen—about 12 light-years out from the center. This ring rotates at 74 miles/sec and is tilted by about 45 degrees to the plane of the galaxy. Furthermore, it contains several subrings. The outermost ring is cool and composed mostly of dust and hydrogen. But as you move to the inner rings the temperature increases rapidly. Near the center it is 10,000 to 20,000 degrees—so high that the hydrogen is fully ionized.

Finally, inside the innermost ring is a region that is void of gas and dust. Nevertheless, the density of stars here is high. (This is where Rieke found several red giants.) Why is it clear of gas and dust? We're not certain, but it appears that an explosion may have scoured it.

Kwok-Yung Lo of the University of Illinois has been studying this region for the last several years. He recently made a detailed radio map of it using the VLA. I asked him about a bar-like segment that some astronomers claim crosses the clear region. "There's a straight section," he said. "But it's not really a bar. The whole cavity inside the ring of whirling matter is filled with streams of ionized gas."

The most intense energy source, both in the radio and infrared regions, is right at the center of the clear region. Astronomers determined some time ago that the mass of the object at the center is high—over a million solar masses. But as the region was studied in more detail the issue became more controversial. Rieke explained the problem to me. "When people looked at the motions of the gas in this region," he said, "they saw that they

Closeup of the very center of our galaxy showing bar and clear region.

were rapid, and concluded that the mass increased very rapidly as you approached the center. Then, when they subtracted out the light of the stars, which was used to give them an idea of the mass distribution, they ended up with a large excess of mass. [This is assumed to be the mass of the central object.] But a couple of groups have now looked at the motion of the stars in this region . . . and that's important because gas is affected by explosions that occur near here, but stars are not. And the velocity of the stars doesn't seem to agree very well with the gas velocity. The speeds are lower. So this means that the mass in the region is less than thought." Rieke also mentioned that his infrared measurements indicate that the stars inside the gas ring are not as densely packed as previously thought.

According to Lo the mass range for the object at the center is between a few hundred solar masses and a million. "But even if it's a few hundred," he said, "it's still not a stellar object. It's still unusual."

So what is the central object? The most promising candidate, as you might expect, is a black hole. But there is controversy. Lo and his colleagues have shown that the radio source at the center is exceedingly small. "It's only about the size of the solar system," he said. This seems to be evidence in favor of a black hole.

Lo says it would be exciting if it were a black hole. "But I have to be honest and say that there's no clear-cut evidence it is a black hole. Of course, there's no evidence that it isn't. It's just not conclusive at this point."

I asked him about the possibility of it being a cluster of stars, or perhaps a single object of some type. "I think it's unlikely it's a cluster . . . the radio source is much too small. But can it be a star—a neutron star, or something like that? If it's a neutron star, its radio properties are very unusual. The real problem is that we don't know its exact mass. That's the missing link."

FILAMENTS

Another strange feature of the central region is the presence of huge filaments. In 1984 Mark Morris of UCLA and Farhad Yusef-Zadeh and Don Chance of Columbia University, using the VLA, discovered three enormous parallel arcs of gas approximately 10–20 light-years thick. They are over 150 light-years long and project out from the plane of the disk.

Astronomers were astounded. Studies soon showed that arcs of this type had to be composed of high-speed particles trapped by extremely strong magnetic fields. They are, in a sense, much like the giant prominences (streams of gas) on the surface of the sun.

Were they being produced by a black hole at the center? In many respects they looked like streams of hot, radio-emitting gas that had been ejected from the core. But at this time we still do not know what causes them.

Soon after these filaments were discovered, much larger filaments were discovered by a Japanese team of radio astronomers of the University of Tokyo's Radio Observatory. They are horseshoe-shaped, and rise about 700 light-years above the galactic plane. They resemble the giant arches of gas that are sometimes seen on the sun, but they are, of course, billions of times larger. It is believed that they are high-speed particles trapped in magnetic fields.

The core of our galaxy is obviously a fascinating place. And although there is still some controversy as to whether a black hole resides there, it is, nevertheless, a site of tremendous power and violence.

CHAPTER 8

A Detailed Look at a Nearby Exploding Galaxy: Centaurus A

The first Australian radio observatory was set up in the Sydney suburb of Dover Heights on a clifftop south of the entrance to Sydney Harbour. Much of the initial work at the site was devoted to the sun. But John Bolton and his colleagues soon became interested in radio sources beyond the sun. One of the first to be studied was Cygnus A. In November 1947, however, several other sources were found, including one in the constellation Centaurus (later called Centaurus A). Bolton decided that a move to a site in New Zealand might be helpful in allowing them to get a better fix on these new sources. So in early spring of 1948 Bolton and his crew moved their equipment to a site north of Auckland. Within a few months they had accumulated considerable data. Bolton then returned to Sydney to reduce it, and within a short time he had identified three sources. One in Taurus was identified with the Crab nebula, one in Virgo with M87, and finally the one in Centaurus with the strange elliptical galaxy NGC 5128.

Bolton remained at Dover Heights for several years directing the building of several larger radio telescopes. Then in 1955 he was invited to set up a radio observatory at Owens Valley in California and he accepted. By this time, however, the Australians had already begun planning a large steerable telescope similar to the one at Jodrell Bank. It was to be built at Parkes.

In 1961 the 210-foot Parkes dish was completed and Bolton returned from the United States to direct the new facility. Centaurus A had now been studied extensively by small radio telescopes. But with the new Parkes telescope it could be looked at in much more detail. One of the first to study it was an American, Marcus Price, who had been an undergraduate at Colorado State University while the Parkes dish was being built. Just before he was to graduate, a physics professor suggested he apply for a Fulbright Scholarship and travel to another country to broaden his experience. Price mentioned the suggestion to another professor who had just read about the progress the Australians were making in radio astronomy. "Why don't you go to Australia," he said. "You could learn a lot about radio astronomy." Price thought about it and decided to apply.

While waiting for a reply from the Fulbright committee he wrote to Joseph Pawsey, the director of the Australia group about joining them. "We'd be glad to have you," Pawsey wrote back. Price was delighted; he notified the Fulbright committee of Pawsey's interest, then proceeded to make plans to leave. Then he got a shock. He got a letter from the committee saying that they had never heard of Pawsey and his group, and they didn't think the trip to Australia was a good idea. Price quickly got in touch with the committee and was able to convince them that the trip would be a valuable experience. He got his scholarship and was soon off to Australia.

By now it had been discovered that the radiation from Centaurus A was polarized. This meant that the radio signal vibrated in a particular direction. If, for example, you tie one end of a rope to a doorknob, then stand back and move the other end up and down, a wave will move along the rope in a vertical direction. This wave is polarized in the vertical direction. In the same way, the radio waves from Centaurus were polarized in a particular direction. Once polarization was discovered, the next step was to search for a possible rotation of the direction of polarization that would be caused by magnetic fields between the Earth and Centaurus A. This is called Faraday rotation. The identification of Faraday rotation was particularly important because it would verify the existence of magnetic fields in Centaurus A and tell us how they were oriented.

Shortly after Price arrived at Parkes the observatory closed for the Easter holiday. As a graduate student, though, Price was required to stay on site to "change fuses and do all the other things you would expect a graduate student to do." Price thought that since he had to be there anyway, he might as well do something useful. The polarization of Centaurus A had just been identified at a wavelength of 11 centimeters. He decided to take a measurement of it at a different wavelength. If it was also polarized at this wavelength it would indicate Faraday rotation. But there was a problem: He was the only one at the observatory and one of the rules was that whenever the telescope was being used, two people had to be in attendance. So he went to the site manager and asked him if he would visit for a while and keep him company. The manager agreed, so while he sat and read, Price made a measurement at a wavelength of 21 centimeters. And, as he expected, there was polarization.

Price was excited about the result and wanted to tell everyone. But there was another problem: Even though a second person was with him during the observation, he really wasn't supposed to be using the telescope. Nevertheless, he casually mentioned it to Bolton, who agreed that it probably was Faraday rotation. Then Price started looking through some of the books on the subject, and he found a prediction by a Soviet scientist that Faraday rotation would not be observable in galaxies. He was disappointed; if true, it meant that what he had observed must have been due to something else.

But was the Soviet scientist correct? Price soon convinced himself that he wasn't, so, along with a colleague, he retook the measurements a few weeks later. They checked the effect at several different frequencies, and in each case there was polarization. It was, indeed, Faraday rotation. Magnetic fields were therefore present in Centaurus A and the radiation (at least some of it) was likely synchrotron radiation.

Centaurus A showing dust lane across it. (National Optical Astronomy Observatories, Tucson, Arizona)

THE RADIO AND OPTICAL IMAGES

Centaurus A has now been studied in considerable detail, and we have an excellent radio map of it. It is one of the strongest radio sources in the sky, not because it is so much stronger than most other active galaxies, but because it is so close. It is only 15 million light-years away, which is only 7 times as far away as the Andromeda galaxy, our closest galactic neighbor.

When the first radio maps of Centaurus A were made, astronomers were amazed by its size. It covered almost 3 million light-years, which is about 10 degrees in the sky, and at that time it was the largest known object in the universe.

There are two huge radio lobes on either side of the optical image that stretch out in the shape of an S. A close look at the overall image shows that it is quite complex. Interestingly, there is considerable asymmetry; on one side, for example, about 70,000 to 130,000 light-years from the center, there is a roughly circular lobe. Nothing similar is found on the opposite side. This middle lobe, as it is called, coincides roughly with a position where filaments are seen in the optical image.

Closer to the nucleus, about 30,000 light-years out, are a pair of smaller radio lobes. They are symmetric about the center and are referred to as the inner lobes. They are about 3500 light-years in diameter, and there is no visible light associated with them. Finally, at the very center is a particularly strong source—presumably the engine that drives the system.

The optical image, known as NGC 5128, is also odd. It appears to be made up of an elliptical and a spiral galaxy. The main component is a bright elliptical galaxy about 45,000 light-years across, with a mass of approximately 300 billion suns. Across its diameter is an intense dark lane, composed of dust and ionized hydrogen. Strangely, although the elliptical portion of the galaxy hardly rotates at all, this dust lane rotates relatively fast.

Another interesting feature seen in the optical image is a series of filaments about 50,000 light-years from the center. They

The central region of Centaurus A illustrating lobes.

appear to be chains of bright young stars. They must have been formed near their present position, as not enough time has elapsed for them to have been formed at the center and ejected to their present position.

THE DISCOVERY OF X-RAY JETS

The lobes are a fascinating feature of the galaxy, but even more amazing is an X-ray jet. It was discovered by Eric Feigelson and Ethan Schreier using data obtained from the X-ray satellite "Einstein."

Feigelson was a graduate student when Einstein was being built. "I met the director of the project, Riccardo Giacconi, when I was an undergraduate," said Feigelson. "I was so impressed I went to work with his group between my junior and senior years."

When he graduated a year later, Feigelson went to Harvard and joined Giaconni's group. The X-ray satellite Einstein was now near completion. "It was a huge project. I worked deep in the bowels of the software for a couple of years writing code, then I participated with the senior scientists in analyzing the data as it flowed in," he said. "It was one of the most exciting experiences of my life. I will be lucky if it will ever be matched again. I got to ride in on my bike each morning and walk into the computer room where last night's data had just been processed. And I would occasionally see an image no human being had ever seen before . . . like the image of Cass A, the supernova remnant. And the image of Centaurus A."

Eric Feigelson.

Feigelson's doctoral thesis, which was titled "X-ray Image Observations of Centaurus A, Virgo A, and NGC 273," was based on the data received from Einstein. It was while working on it that he discovered the X-ray jets in Centaurus A.

Also involved in the discovery was Ethan Schreier, Feigelson's supervisor. Schreier started out in theoretical high energy physics at M.I.T., but when he graduated in 1969 he switched to X-ray astronomy. "I got involved with the first X-ray satellite [Uhuru] by accident," he said. "I got a part time job in astronomy while I was a graduate student and got very excited about observational astronomy." So, upon graduation he began working on the data from Uhuru, most of which was related to binary X-ray sources. "It was the golden age of X-ray astronomy," he said.

He stayed with the group for a decade. Then, when plans were made for a more sophisticated X-ray satellite, Schreier immediately became involved. The satellite was, of course, Einstein—the first X-ray satellite to give direct images. He decided it would be interesting to look at the X-ray images of some of the extragalactic sources—particularly Centaurus A. And when he did, he (along with Feigelson) discovered the jet.

I asked him what his reaction was when he first saw the jet. "We thought it was a mistake," he said. "But it was very exciting because it really did look like a jet. We were worried, though, that instrument effects could have caused it . . . that the satellite could have moved a little and blurred the image."

Feigelson's reaction was similar. "We never expected it. But it really wasn't an individual discovery, as in the old days when a guy with a telescope discovers something. This was a project involving hundreds of millions of dollars and hundreds of people. And Centaurus A was a very obvious and prominent source. Still, no one predicted it would contain an X-ray jet, so it was a surprising discovery."

Checks were made, of course, and it was soon verified that the satellite had not moved. The image was, indeed, an X-ray jet.

But if there was an X-ray jet, was there also a radio jet? No one had ever seen one. But Centaurus A had never been looked at with a high-resolution instrument. About this time the VLA (Very Large Array), a group of 26 radio telescopes in New Mexico, was just being completed. Feigelson and Schreier would be able to use the new array to look at Centaurus A.

Schreier sent a preprint of their paper to a radio astronomers at the VLA named Jack Burns. Burns was interested so he got in touch with Schreier, and the three of them teamed up to look at Centaurus A in the radio region.

Burns got his undergraduate degree is astrophysics from the University of Massachusetts in 1974, and his Ph.D. from Indiana University, where he worked on radio galaxies in clusters. Upon graduation he went to the VLA, then later to the University of New Mexico.

By 1979 most of the VLA had been completed, but the northern arm had just gone into operation when the group got organized. The northern arm was needed to do high resolution studies in the south, and Centaurus A was just barely above the southern horizon. (Its maximum angle above the horizon is only about 14 degrees.) So Burns, Feigelson, and Schreier looked at Centaurus A—in particular, at the region where the X-ray jet was—and, sure enough, there was a radio jet.

"We were lucky," said Feigelson. "If Centaurus A had been in the northern hemisphere and easily observed, astronomers would have discovered the radio jet years ago. It just happened to be located in a region of the sky that was inaccessible to high-resolution radio telescopes during the 1960s and 1970s. So we were lucky to discover it first." He laughed. "But I don't mind. It was a very exciting time . . . and a nice piece of work."

Also reflecting on the discovery, Schreier said, "When you get a new observing technique you discover things . . . whether or not you see them in advance, and one thing leads to another. We discovered the X-ray jet with the first X-ray imaging telescope and we were able to follow it up with the discovery of the radio jet using the first radio telescope that was able to resolve it. That's why it's always exciting to be associated with new projects. There are always things that are waiting to be discovered."

The jet appears to start near the core. It extends out to one of the inner lobes, about 20,000 light-years away. Since the X-ray and radio jets are in the same position, they are, in effect, the same jet—a stream of high-energy electrons giving off both X rays and radio waves. The electrons are spiraling around magnetic field lines giving off synchrotron radiation. Both the X rays and the radio waves, in fact, appeared to be synchrotron radiation.

It was later discovered that there is a tiny jet—only about 4 light-years long—extremely close to the core. It points in the same direction as the larger jet and is no doubt just its inner section.

There are many knots along the length of the jet. They may play an important role in reenergizing the beam. A simple

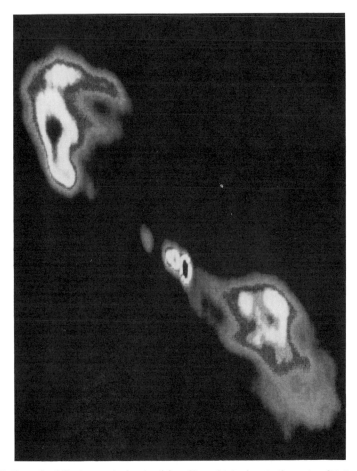

Radiograph of Centaurus A showing lobes. X-ray jet is close to the center. [National Radio Astronomy Observatory (NRAO) and observers J. O. Burns, E. J. Schreier, and E. D. Feigelson]

SYNCHROTRON RADIATION

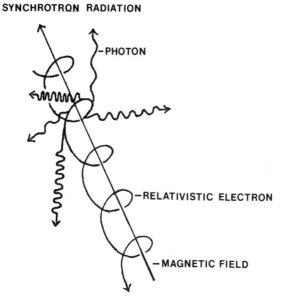

How synchrotron radiation is produced. Electron is spiraling around magnetic field lines.

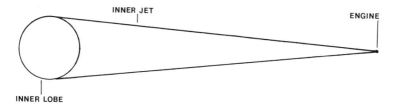

Schematic of inner jet and "engine" at the center of Centaurus A.

calculation shows that there is a serious problem in relation to the energy of the jet. It is assumed that both the radio waves and the X rays are generated via synchrotron radiation. In other words, they are both generated as high-speed electrons spiral along magnetic field lines. But if these electrons produce X rays they're going to lose energy rapidly. There is no way they could continue to produce energy over 20,000 light-years—the length of the jet. And as they lose energy they have to be reenergized. In fact, they would have to be reenergized several times along the length of the jet, and there is evidence that they are being reenergized at the knots. A shock wave may occur when the beam hits some intergalactic material which may speed up the electrons so that they can radiate again. Later on, when they slow down, another shock wave may speed them up again.

"Some of the superconducting models that we've done recently suggest that such knots occur naturally from shocks," said Burns. It is also possible, though, that the energy associated with the turbulence in the beam is sufficient to accelerate the electrons. And it is possible that protons within the beam collide with the electrons and reenergize them. We don't see the protons because they do not produce significant radiation. Most astronomers, however, are convinced that the shock wave mechanism is the most likely explanation.

But is the energy produced by synchrotron radiation? Feigelson says he is no longer sure. "It was an amazing discovery, but on a more detailed level it has proven to be a little exasperating," he said. "Of the various mechanisms for producing the energy, we assumed it had to be synchrotron radiation. To be frank, since then, I've backed away a bit. The data are not that conclusive. Astronomically the experiment was a success: We saw something in the sky we had not known. Astrophysically, in other words, understanding the process, it's been up in the air. And in the 8 years since we haven't been able to completely explain it."

ASYMMETRY OF THE JET

One of the strangest things about the jet is that it appears on only one side of the galaxy. It is possible, of course, that another jet exists on the other side that we can't see. Jack Burns, however, is skeptical. "We've looked pretty close," he said. "And our data show no evidence whatsoever of another jet. It's either not there or it's completely invisible. If this is the case the mechanism that is illuminating it on the other side is quite different from the one on this side."

A number of scientists have argued that it may be undetected because of relativistic effects. We talked about this in Chapter 5. Ethan Schreier, however, does not believe that these arguments apply in the case of Centaurus A. "We know too much about Centaurus A," he said. "There are too many constraints, and that kind of model starts becoming very hard to sustain. I feel there's only one jet going in one direction."

Another explanation has been proposed by Lawrence Rudnick of the University of Minnesota. He suggests that the jet switches sides every so often. But again there are many difficulties and most astronomers don't accept the idea.

"It's just another mystery we don't know the answer to." said Burns. "Most low luminosity sources like Centaurus A tend to have two jets. I guess Centaurus A is just an exception."

AT THE CORE: THE BLACK HOLE

As in the case of other active galaxies, we're faced with the problem: What is supplying the energy that is producing the lobes and jets? Obviously, there has to be an "engine" of some sort at the center. Furthermore, it has to be extremely small, as observations of X rays and radio waves show variations of less than 24 hours, indicating that the engine can't be over a few light-days across.

What is it? Most astronomers, as you might have guessed, are convinced it is a black hole. The most efficient way of extracting energy from matter is to have the matter fall into an extremely strong gravitational field. And the best source of such a field is a massive black hole. In the case of Centaurus A, this black hole would have to have about a billion solar masses.

Furthermore, the black hole is likely spinning so that any matter falling toward it would be drawn into an accretion disk around it. It would also likely have a magnetic field, making it possible to accelerate particles into a highly collimated beam—in other words, a jet.

I asked Burns how convinced he is that there is, indeed, a black hole at the center. "As convinced as I can be without directly seeing it," he replied. "No one has presented a replacement model that allows us to get the amount of energy at radio and other wavelengths out of a galaxy as the black hole model does. Black holes are the most efficient way of turning mass into energy."

Schreier replied to the same question as follows. "There has to be an energy source. And in terms of active nuclei the black hole is one of the best. To the extent that black holes are a reasonable way of explaining energy, they're right up there." He chuckled. "If you like black holes, they're a great candidate. I neither like nor dislike them."

CANNIBALISM

Of course, if there is a black hole at the center, there's still the question of fuel. What is fueling it? One possibility is nearby galaxies. A survey of nearby clusters shows that a large elliptical galaxy frequently lies at the center of the cluster. And in many cases the elliptical is active. Why? Many astronomers believe that they have become large as a result of numerous encounters. When two galaxies pass near to one another gravitational forces pull material from one onto the other. Repeated encounters will frequently create a giant. And once it is a giant it will no doubt begin to cannibalize its neighbors. One by one, it will "eat" them, and in the process it will become active. Is this what is happening in the case of Centaurus A? As early as the 1950s, Walter Baade and Rudolph Minkowski of Mount Wilson and Palomar observatories suggested that the form of NGC 5128 indicated a collision between a spiral and an elliptical galaxy. Computer simulations of such a collision by Allan Tubbs of the National Radio Astronomy Observatory and others have, in fact, indicated that a collision between a spiral and an elliptical would result in a shape similar to that of NGC 5128.

Looking carefully at NGC 5128 we see that the overall galaxy does indeed look very much like an elliptical. It has almost no rotation and little gas and dust, which are characteristic of ellipticals. The dark central band, on the other hand, has properties that are normally associated with spirals. It contains considerable dust, gas, and neutral hydrogen; furthermore, it's rotating relatively fast, and there's evidence of star formation.

The evidence for collision appears to be strong. Is it likely, then, that cannibalism is taking place? Unfortunately, the idea poses a serious problem. There are no other galaxies within a few million light-years of NGC 5128. Most active ellipticals of this type have numerous other galaxies around them.

"It's a peculiar system," said Burns. "If it were in the center of a group or cluster I would be inclined to go along with cannibalism. But it's so isolated. You would have to say it has eaten all its neighbors, and the likelihood of this happening is not high." Burns went on to explain that some of the neighbors, assuming it originally had some, would move faster than others, and they would tend to escape the cannibalism. The time between collisions in clusters is very long—of the order of the age of the universe. If Centaurus A started off with a cluster around it, it seems unlikely that it would have eaten them all. Not enough time has passed.

THE FUTURE

What is planned in the way of future studies of Centaurus A? One of the most exciting events will be the launching of the German satellite ROSAT in the spring of 1990. Several U.S. projects will be aboard. One, it is hoped, will be that of Ethan Schreier and Eric Feigelson. They have submitted a proposal to do some high resolution studies of Centaurus A. "We would like to get a spectrum of the jet and the X rays so that we can resolve some of the problems that are still outstanding," said Feigelson. But even if these problems aren't solved with ROSAT, they should be when the X-ray satellite AXAF is put in orbit using the space shuttle in 1997.

Aside from satellites, considerable other work is planned. Modeling of the basic structure of Centaurus A is now being done using large Cray supercomputers. Infrared astronomers are also getting into the act, studying the infrared radiation from the core. And several projects involving Centaurus A are planned for the Hubble space telescope.

"We're also planning a whole battery of new observations with the VLA," said Burns. "We're now capable of imaging larger fields of view, and would like to look at the radio emission that exists between the inner and middle lobes." Optical images have also indicated that there may be young stars in some of the emission knots. Burns is looking forward to studying them.

CHAPTER 9

Colliding Galaxies: The Discovery

We have seen that some galaxies are extremely energetic, their energy being produced by an "engine"—perhaps a black hole — at their center. This energy manifests itself in the form of huge radio lobes on either side of the galaxy, some so large that they cover millions of light-years. Powerful jets are also visible in a few cases. But all radio sources in the sky are not of this type. In some cases the radiation is produced by the interaction of two galaxies. We saw earlier that Cygnus A was initially thought to be two galaxies in collision. And although we now know this isn't the case, we also know that there are galaxies that are interacting, and in some cases actually colliding and merging.

What is the probability of a collision? It's much greater than you might think. If you compare the space between galaxies to their size you find that, on the average, they are separated by about 10 to 100 times their diameter. Think of a crowd of, say, 50 people each separated by only 10 times their diameter, then have them move at random. I think you'll agree that there would be a large number of collisions. And, of course, you would get the same thing in a crowd, or cluster, of galaxies separated by only 10 times their diameter, assuming they were moving at random.

What would happen if two galaxies did collide? To answer this we have to look at the distance between the stars within a galaxy. It is known that stars are separated by approximately 100 million times their diameter. If we assume that they are moving

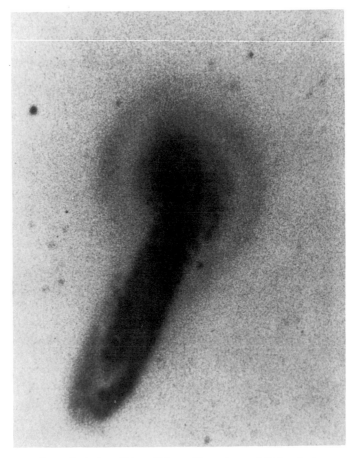

Two galaxies in collision. (Palomar Observatory and Halton Arp)

through space at roughly the same speed that galaxies move, we find that a star is about a trillion times less likely to collide with another star than a galaxy is with another galaxy. What does this mean? Strangely enough, it means that if two galaxies collide, the stars within them will just glide past one another without any physical contact whatsoever. It sounds crazy, but it's true.

Is there any evidence that galaxies do, indeed, collide? Interestingly, the first spiral galaxy ever identified appeared to be in the act of collision. About 1850 Lord Rosse of Ireland looked through his 72-inch reflector at the 51st object in Messier's catalogue and was surprised by its appearance. It looked like a whirlpool. He had no idea what the object was, but he did notice that it was interacting with a neighbor. He wrote, "The connection of the companion with the greater nebulae . . . adds . . . to the difficulty of forming any conceivable hypothesis [as to what the object might be]."

An even more striking system of two interacting galaxies was found by C. O. Lampland of Lowell Observatory in 1917. In this case, not only were the galaxies touching one another, but two long luminous filaments were protruding from them. They looked so much like the antennae of an insect that they were eventually called the Antennae. Lampland also had no idea what was going on, as the concept of a nebulae as an "island universe of stars" had not yet been formulated. It didn't come until the late 1920s.

Are interacting galaxies of this type common? A casual look at an atlas of galaxies would seem to indicate that they are rare. But astronomers have found that about 1% to 2% of all galaxies appear disturbed in some way. This usually occurs when they come in pairs, so it is reasonable to assume that they are perturbing one another. Luminous filaments or bridges are, in fact, frequently seen between them.

Incidentally, when astronomers first found galaxies of this type they thought they looked so peculiar that they referred to them as peculiar galaxies, and the name has stuck.

What causes the filaments and bridges associated with these peculiar galaxies? Although there is still some controversy, most astronomers are now convinced that they are created by tidal forces. As the name suggests, these are the forces that create tides here on Earth. The high tides caused by the moon result from the moon's gravitational pull on the near side of the earth being considerably stronger than that on the far side. The ocean on the near side is therefore pulled toward the moon with a greater force, and it bulges out in this direction. In fact, not just the ocean bulges out; land masses are also moved in the direction of the moon by several inches.

The important feature is not the overall strength of the gravitational field, but the "difference" in gravitational pull between one point and another. If this difference is large (i.e., if the tidal force is large) each of the objects will be stretched, or pulled apart, in the direction of the other object.

When two galaxies approach one another there will obviously be huge tidal forces between them, and these forces will tend to stretch the two galaxies. In other words, the near sides of the galaxies will be pulled toward one another with a far greater strength than their far sides, and their shape will therefore become distorted. It is important to note that two galaxies can interact, and considerable distortion can occur, even if there is no direct collision.

In a direct collision the distortion, and the damage to the two galaxies will likely be considerably greater. But, as I mentioned earlier, the stars themselves will not collide. Let's look at this case in detail. What happens? We know, first of all, that galaxies are surrounded by a huge halo of "dark matter." We can't see or photograph this dark matter but we know it's there because of the way the galaxy spins, and how it affects stars that stray above and below its main disk. Astronomers are still not sure what this dark matter is. But they are relatively certain it exists.

Besides the dark matter in the outer halo, there is also dark matter throughout the galaxy. Furthermore, there is a considerable amount of gas and dust scattered throughout it. It is this gas, dust, and dark matter that will cause most of the fireworks when the collision takes place.

Suppose, then, that the two galaxies are approaching one another. The first thing to interact will be the two halos. What will happen when this occurs? That depends on whether the galaxies are spinning in the same direction or in opposite directions. If they are spinning in opposite directions their motion will be halted as their halos begin to interact. They will then move back from one another. But eventually, because of the gravitational attraction between them, they will begin to move together again.

And although the individual stars will not collide, they will feel the overall gravitational pull of the other galaxy and the resulting tidal force will perturb many of them from their orbit. In some cases they will be forced into an orbit closer to the center of the galaxy; in other cases they will be forced into orbits farther out. The effects of this will be significant. We know that the stars in a galaxy obey the same laws that the planets in our solar system do. This means that if a star is perturbed to an orbit closer to the center, it will speed up, just as a planet in the solar system would if it were closer to the sun. Actually, the effect is not exactly the same, since a galaxy's mass is spread out. In the case of the solar system, almost all of the mass is in the sun. Nevertheless, there will be a change in speed when a star changes orbit. And as a result, as the star continues to move, the overall galaxy will begin to change shape.

The real fireworks, however, will come from the gas and dust between the stars of the galaxy. It will collide with the gas in the other galaxy, producing a tremendous amount of radiation. From the Earth we will see the object as a radio source.

EARLY MODELS

If we know how massive the two galaxies are, how many stars they contain, how fast they are approaching one another, and how they are spinning, we should be able to calculate exactly what will happen when they collide. The perturbation of each star depends only on the gravitational field it feels, and astronomers know how to calculate this force. Erik Holmberg of the Lund Observatory in Sweden became interested in this problem in 1940. He realized that tidal forces would be an important factor in a collision, and that stars would be perturbed from their orbits. It was obvious to him, however, that a large number of detailed calculations would be needed. After all, there are about a hundred billion stars in each galaxy, and the gravitational field that each of these stars experience would have to be determined. Computers were not available at that time, however, and Holmberg knew he could not make such detailed calculations. He therefore did the next best thing: He built a physical model to represent the collision.

He realized that the gravitational field fell off as the inverse square. In other words, the field of a star gets weaker as you move away from it, and the decrease is proportional to one divided by the distance squared. It turns out that intensity of light falls off in exactly the same way. For example, if you double your distance from a source of light, it will appear only one-quarter as bright.

Holmberg therefore set up arrays of light bulbs to represent the gravitational fields of the stars in a galaxy. He needed detectors to measure the amount of light, so he placed photocells next to each of the bulbs. He then moved the arrays toward one another; individual bulbs were moved according to the "imbalance" in the net light they received. This represented the tidal force at the point of the star. He kept track of the paths of each of the bulbs as the two arrays moved through one another. And, as expected, he found that there were strong tidal forces between the two systems. In particular, he saw that galaxies would be disturbed both on the near side and the far side. We see the same phenomenon with tides here on Earth. There is a high tide in the direction of the moon, but there is also one on the opposite side of the earth, away from the moon. You can think of the high tide on the opposite side of the earth as resulting from the gravitational pull being weakest there, and so the water is, in effect, left behind.

Holmberg was not, however, able to show how long narrow filaments, such as those in the Antennae, were produced. He published the results of his experiment in the *Astrophysical Journal* in 1941. And although his approach was entirely new, it generated little interest among astronomers. The idea of colliding galaxies was just too far outside the mainstream. Ten years therefore passed before others became interested in the problem.

Interest was not sparked again until 1951, when Walter Baade and Rudolph Minkowski, using coordinates sent to them by Graham Smith of Cambridge, found an optical counterpart to the radio galaxy Cygnus A. The object amazed them: it appeared to be two galaxies in collision. They made a calculation to see if the probability of such collisions was very high. And they were surprised to find that it was. They found, for example, that in the Coma cluster, a group of about 500 galaxies in a space 2.6 million light-years across, at least two collisions should be occurring at any time.

The excitement generated by the Baade and Minkowski discovery caused astronomers to begin looking for other examples of collisions and soon many were found.

THE PALOMAR–NATIONAL GEOGRAPHIC SKY SURVEY

A major problem, however, faced astronomers in their search for evidence of collisions. In most searches large reflectors such as the 100-inch Hooker reflector at Mount Wilson were used, and these instruments were not adequate for photographing filaments and bridges between galaxies. Appendages of this type are so dim that exposures of many hours—in some cases literally all night—were needed to record them.

When the 200-inch Palomar reflector went into operation exposure times could be shortened, but there was another problem. The field of view was so small that few new systems were discovered. In 1949, however, another telescope went into operation at Palomar—one of quite different design. Called a Schmidt telescope, it used both a mirror and a lens to gather light, and consequently it was much faster than the 200-inch reflector. It was therefore ideal for photographing galaxies with

The 48-inch Schmidt telescope at Palomar. (Hale Observatories)

dim filaments. Furthermore, it had an exceedingly wide field of view.

The first project of the 48-inch Schmidt was a sky survey cosponsored by the National Geographic Society in Washington. The entire northern hemisphere available to the tele-

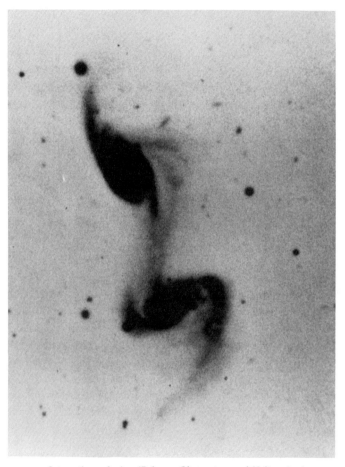

Interacting galaxies. (Palomar Observatory and Halton Arp)

scope was photographed over a period of about 7 years. Approximately 1700 plates were obtained, and it was soon evident that many of them showed interacting galaxies.

It was only a matter of time before someone started paying attention to them. And shortly after the plates were published the Soviet astronomer Boris Vorontsov-Velyaminov of Sternberg

Astronomical Institute of Moscow began scanning them, looking for interacting galaxies. By 1959 he had found 355. Using them, along with a few from other sources, he published an atlas of interacting galaxies. He extended it in 1964 with the publication of a second atlas.

Vorontsov-Velyaminov listed the NGC (New General Catalog) numbers of the objects, gave their coordinates and magnitudes, and gave a brief description of each. Many of his interacting galaxies had luminous filamentary bridges between them.

He found that the connecting filaments were usually short and broad when associated with elliptical galaxies, but were frequently long and narrow when associated with spirals. Calculations showed that some of these filaments were as long as 200,000 light-years. This is twice the width of our galaxy.

Strangely, Vorontsov-Velyaminov did not believe the filaments were caused by gravitational tides. He calculated how long it would take to produce a filament 100,000 light-years long, and found it to be about 400 million years. His calculations indicated, however, that over such a long time the filament would become quite thick.

He therefore rejected the idea of tidal forces and tried to explain the filaments in terms of gas motion along magnetic field lines. Unfortunately, he did not explain why the gas moved along the field lines, other than suggest that there might be an unknown force causing it.

ZWICKY AND LINDBLAD

While Vorontsov-Velyaminov was collecting photographs for his atlas, Fritz Zwicky at Mount Wilson Observatory began looking for interacting galaxies. Eccentric, cranky, and difficult to work with, Zwicky was a paradox. He was a first-rate astrophysicist: the first to predict the existence of neutron stars, the first to find dark matter in clusters of galaxies, the first to search for supernovae in other galaxies—and the first to find them. Many of his ideas were ingenious, but because he was such a crank most astronomers did not take them seriously. Born in Bulgaria in 1898, Zwicky came to the United States after receiving his Ph.D. in 1925. He had many interests—jet propulsion, rocketry, crystallography. And he eventually accumulated about 50 patents for various devices. But he wasn't exactly everyone's favorite co-worker. He firmly believed that he was superior both mentally and physically to everyone whom he worked with. As a result he looked down on most of their accomplishments. This, as you might expect, backfired, and he was gradually excluded from important meetings and his time on the large telescopes was reduced. Much of his important work was, in fact, done with a rather small 18-inch Schmidt telescope.

Zwicky began to take an interest in interacting galaxies shortly after the Palomar–National Geographic Survey plates were published. He had already spent considerable time working on clusters of galaxies, so the interest was perhaps expected. In a publication in 1956 he reported that several thousand interacting systems—systems showing luminous bridges—had been found on the Palomar–National Geographic plates. He was convinced that they were due to tidal forces, but bothered by the fact that some of them were extremely narrow.

Zwicky published drawings of how he believed galaxies interacted. He also showed how elliptical galaxies could be transformed into spirals in an encounter, and how long filaments might form. He rephotographed many of the interacting galaxies he found on the Schmidt plates using the 200-inch Palomar reflector.

Others shared Zwicky's opinion that the bridges and filaments were produced by tidal forces. The Swedish astronomers Bertil Lindblad and his son did a theoretical investigation of them. The younger Lindblad, in fact, did a detailed computer study of the collision of two galaxies. He began by considering a star in each of two approaching galaxies and tracing their path

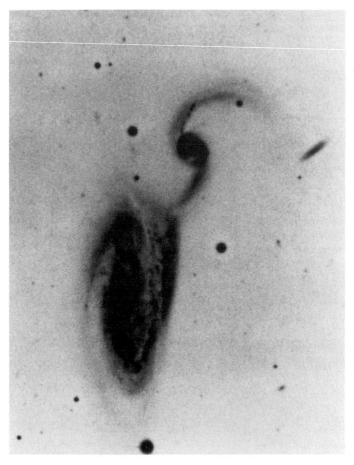

Large galaxy interacting with smaller one. Note long filaments on the smaller one. (Palomar Observatory and Halton Arp)

during and after the interaction. He found that both of the stars would be severely perturbed from their orbit; in some cases they would move in the direction of the perturbing galaxy, in others they would move away. His calculations hinted at connecting links between interacting galaxies, but he found that special conditions were needed to produce the narrow filaments observed.

More complete models of interacting galaxies were made by Jorg Pfleiderer of Germany about 1960. His major interest was spiral arms. He wanted to find out if spiral arms could be produced in the collision of two disk galaxies. Computers were still relatively slow at this time, so he had to make a large number of approximations. He considered each of the galaxies to be made up of a group of stars, and calculated the tidal force on them. But he asssumed that the gravitational attraction came only from a point at the center of the galaxies. We know, of course, that the gravitational field is actually spread throughout the galaxy. He found that he was able to produce curving filaments but they were extremely short-lived. He therefore concluded that spiral arms could not result from collisions. His results were published in 1963.

Although much of the work at this time was theoretical, observational work was progressing. The British astronomers Geoffrey and Margaret Burbidge followed up on Zwicky's work using the 82-inch reflector at McDonald Observatory. They photographed many of the interacting systems and also studied them using the spectroscope. Their conclusion was that these systems were still in the process of forming.

Within a short time, however, it was shown that the object that had initiated most of the interest, namely Cygnus A, was not two galaxies in collision. The radio emission appeared to be coming from two regions on either side of it. Furthermore, it was synchrotron radiation, which occurs when charged particles spiral along magnetic field lines. There was no apparent way this type of radiation could be produced by colliding galaxies. The major problem, however, was that many other radio galaxies were being discovered and they did not look like colliding galaxies. A different mechanism for their energy generation was obviously needed and it was soon found: The core of the galaxy contained a powerful "engine." With this there was no need for collisions, and interest in them soon waned. From the early 1960s through the early 1970s almost no attention was given to colliding galaxies. One important event, however, did occur during this time.

ATLAS OF PECULIAR GALAXIES

In 1966 Halton Arp of Mount Wilson-Palomar Observatory published his *Atlas of Peculiar Galaxies*. One of the first recipients of a Ph.D. in astronomy from Caltech, Arp went to work at the Mount Wilson-Palomar Observatory in 1957. He soon became interested in the work of Zwicky and Vorontsov-Velyaminov. Furthermore, he found that he was not satisfied with the theories of galaxies that were accepted at the time. They didn't seem to explain many of the galactic forms—in particular, peculiar galaxies. In an effort to understand things better he decided to assemble an atlas of peculiar galaxies.

Using Zwicky's work and Vorontsov-Velyaminov's catalog as a guide he began photographing peculiar galaxies using the 200-inch reflector. By 1966 he had photographed 338 systems which he assembled into an atlas and published. The resolution of these photographs was considerably better than those in Vorontsov-Velyaminov's atlas. Furthermore, Vorontsov-Velyaminov's atlas had limited circulation, so, for most astronomers, Arp's atlas was their first glimpse of the phenomenon of colliding and interacting galaxies.

Unlike Zwicky, however, Arp did not believe that the appendages were produced by tidal forces. His explanation was more in line with Vorontsov-Velyaminov's. He believed that new forces, or perhaps forces that had not been previously considered may be responsible for them. He also felt that magnetic fields and ionized gas played a role. Today, however, most astronomers feel that magnetic fields play a minor role, if any.

TOOMRE AND TOOMRE

From 1960 to about 1970 there was little progress in understanding colliding galaxies. Quasars were discovered and soon

Alar Toomre.

astronomers found that both active galaxies and quasars were being powered via an "engine" at their core. Still, there were objects in the sky—for example, the Antennae—that appeared to be galaxies in collision. Alar and Juri Toomre became interested in these objects and decided to try to simulate them, using computers.

Born in Estonia, the Toomre brothers were left homeless after the war. They were refugees, wandering from place to place with their parents. Finally, their father found someone in Ohio who would sponsor them as emigrants to the United States. And in October 1949, they set sail, arriving in New York Harbor late in the month. Alar was 13 at the time, and Juri 10.

They lived and worked on a farm in Ohio for a while, then moved to Long Island. "We weren't really country folk, so we got away from the farm as soon as we could," said Alar.

Four years later Alar was a freshman at MIT, where he majored in aeronautical engineering and physics. From there he went to Manchester University in England and a few years later he obtained a Ph.D. in fluid dynamics. Upon returning to the United States he joined the applied mathematics group at MIT, where he has been ever since.

Juri followed in his older brother's footsteps. He graduated from MIT in aeronautical engineering, and went to Cambridge in England where he also obtained a Ph.D. in fluid dynamics. Then, after a brief stay at New York University, he went to the University of Colorado, where he is today.

I asked Alar how he got interested in colliding galaxies. "In the late 1960s Chris Hunter, a fellow associate professor at MIT, and I became interested in the 'warp' [a small twist in the disk] that had been discovered in the mid-1950s in several galaxies, including our own. We wondered how these warps had come about . . . particularly the one in our galaxy. People had already speculated that maybe the Magellanic Clouds had come by and bent it. So we decided to look into this idea. . . . I guess desperate men try desperate things." He laughed, then continued. "We blissfully went along thinking about warping for a year or so, then finally in 1969 I had the good sense to do some test particle calculations to see how much damage would be done. And that was a real stunner . . . . there were terrible distortions in the horizontal plane . . . . our model became a beautiful spiral. But it didn't look at all like our galaxy. That dashed our dreams."

But the fact that the model galaxy was severely distorted excited Alar. He thought it was worth following up on. About this time his brother Juri returned from England and began working in the Department of Mathematics at New York University, and at the Goddard Institute for Space Studies. Alar suggested that they work together on a project, since they were

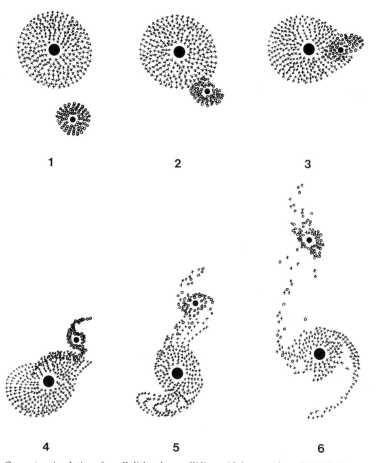

1 2 3

4 5 6

Computer simulation of small disk galaxy colliding with larger galaxy (M51) by Toomre and Toomre.

both in the same area. "Brothers ought to do something together at some point," he said. They talked about several projects, then decided on a study of colliding galaxies. They had access to the large computers at the Goddard Institute and also its excellent plotting facilities.

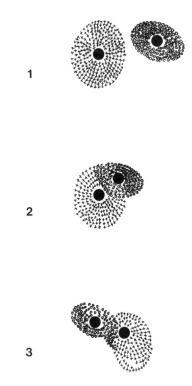

1

2

3

Computer simulation of "Antennae" collision by the Toomres (continued on next page).

Alar said that he was quite aware of the work that Baade and Minkowski had done. "I knew they had been laughed at by the hot-shot theorists. 'Ha, ha . . . don't these two observers know that when galaxies collide they can't possibly do anything.' Of course Baade and Minkowski were wrong in some of the things they said. But isn't it funny how the wheel of fortune has turned almost full cycle. I had the feeling at the time that these two old boys had got too rough a treatment from the theorists. One couldn't help but feel a little sympathy for them."

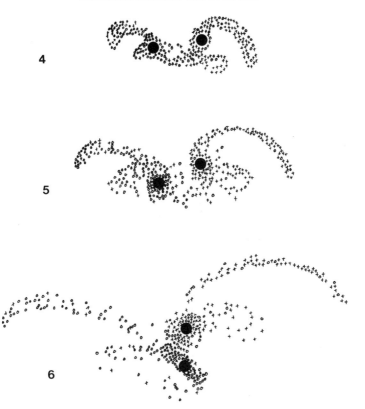

Working with the computers and plotters at Goddard, the Toomres therefore began writing programs that would simulate the collision of two galaxies. They were particularly interested in seeing whether tidal forces could produce the filaments and bridges seen in many of the interacting systems. A galaxy is, of course, made up of billions of stars, and they realized that they would not be able to simulate it exactly. So they approximated.

"We made the rather rash and extreme approximation, for reasons of efficiency, that all the mass of the galaxy was in the middle of it. We did that with both galaxies. Then it was a simple integration over all the lesser particles that chose to fly along with one or the other of the galaxies. It's remarkable how far this ridiculously oversimplified approximation actually describes the probable truth. When another galaxy goes by it applies such a severe and sudden force that one doesn't need to be subtle. The impulse from the neighbor largely does it."

In their first studies they assumed one of the galaxies was made up of 120 test points (stars) while the other was assumed only to be a single center of mass. From the very first runs they were amazed. In addition to considerable distortion on the near side, they saw that the far side of the galaxy was also considerably distorted. "We were delighted to see the counter-arms on the far side," said Alar. "Tides by their very nature are two-sided, so we expected something. Still, it was dramatic. It was a pleasant scientific discovery to find how easy and natural it was to make relatively narrow curving sickles on the far side. This has come through loud and clear in more recent work; it's now been demonstrated umpteen times. But we were the first to demonstrate it."

After a number of successful simple simulations, they set out to see if they could produce several of the well-known collisions in the sky: the Antennae, the Mice, and the Whirlpool galaxy. Each of their model galaxies now consisted of an array of approximately 300 test particles. After many trials—in some cases, hundreds—they were able to reproduce all of the above collisions. In particular, they were able to show that narrow filaments and bridges could be produced via tidal forces. "The Antennae gave us a lot of pleasure," said Alar. "Until we reproduced something like it, we had been concerned whether it was possible to obtain crossed tails."

They found that a number of conditions were required if the filaments and bridges were to be narrow. First of all, the two galaxies had to approach one another at a relatively slow speed. And they had to penetrate one another, but not too deeply. Furthermore, the approaching galaxy had to approach in rough-

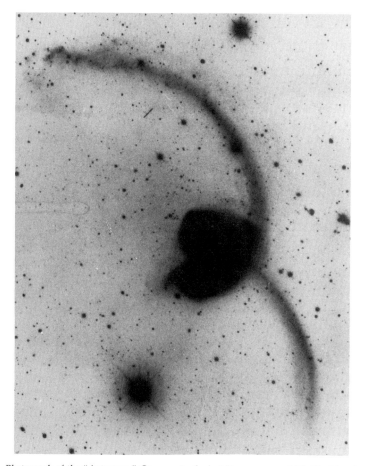

Photograph of the "Antennae." Compare to the last figure on page 187. (F. Schweizer)

ly the same direction as the other one spun. They found, in fact, that spin direction was particularly important in the simulations. Two galaxies approaching one another with opposite spins would react quite differently from two approaching with the same spins.

Somewhat surprising were the delayed effects. The two galaxies only felt one another's gravitational field for a relatively short period of time when they were close together. But during this time many stars would be perturbed to orbits farther out and closer in. As these stars continued in their orbits they moved at different speeds.

I asked Alar what the immediate reaction to their work was. "I wish I could say I was a voice in the wilderness, appreciated only later," said Alar. "But the reaction was quite the opposite. There was an outpouring of nice comments almost immediately. Of course it wasn't just us doing these things. Several others had by then treated the same problem."

Alar is no longer active in the field, although he says that he closely follows its progress. "I pay a lot of attention to the papers, but I'm not particularly active any more because I like to do things where I make a difference. And colliding galaxies have become extremely popular in recent years. Many people are working in the area." He was, however, chief organizer of a conference in Heidelberg in 1989 on the dynamics of interacting galaxies. He said that he was particularly pleased by the attendance; there were over 200 participants. Almost everyone who had done anything in the area was there.

Interest in colliding galaxies is increasing rapidly. But most intriguing, perhaps, is the merger of two galaxies. What happens when two galaxies merge? We will look into this in the next chapter.

CHAPTER 10

Merging Galaxies

Let's take a closer look at the collision process. There are obviously several things that can happen in a collision. First of all, the two galaxies need not penetrate one another; if they pass close enough, tidal forces will disrupt the stars and cause considerable damage. Interestingly, the most damage is done at relatively low speeds. If they pass one another fast enough, few, if any, of the stars will be disturbed. If they penetrate one another, on the other hand, the damage in most cases will be considerable. And, of course, it will be greater yet if they collide head-on.

What, in fact, would happen in a head-on collision? One possibility, discussed earlier, is that they would pass right through one another (usually with some damage to both). But it is also possible that they might merge, and if they do they will end up as a single galaxy. This is the case that will concern us in this chapter.

Let's begin with the conditions that are needed for a merger. Speed is, of course, critical. If it is over about 600 kilometers per second, the galaxies will pass right through one another. If it is only, say, 200 kilometers per second, the galaxies will merge.

But if they do merge, what will they end up as? To answer this we have to look at the various types of galaxies that astronomers observe. As we saw in an earlier chapter, there are two main types: spirals and ellipticals. As their name suggests, ellipticals are oval in shape. They are classified according to their

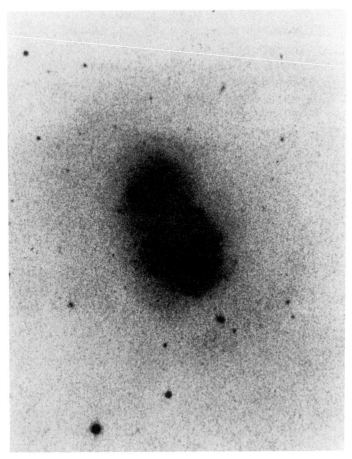

Two galaxies in the process of merging. (Palomar Observatory and Halton Arp)

appearance in the sky, which may or may not be their actual shape. A pancake-shaped elliptical that is seen edge-on, for example, will appear cigar-shaped.

The stars in an elliptical galaxy are generally old, and there is no gas or dust in them. This seems to indicate that they are old galaxies, but as we will see, this may not be the case. Spirals, on the other hand, appear to be young. They are disk-shaped, with long trailing spiral arms. They have considerable gas in their arms and new stars are being formed here. The core, or hub, of a spiral, on the other hand, looks like an elliptical. In most cases it contains no gas, and consists of old stars.

But are spirals actually younger than ellipticals? Most astronomers do not believe this is the case. In fact, for years, the traditional view has been that both types of galaxies formed a few hundred million years after the big bang. Elliptical galaxies presumably formed out of nonrotating gas clouds. Stars formed rapidly, with the most massive ones soon exploding as supernovae. Smaller stars formed from the debris, and again the largest of them exploded, until finally there was nothing left but small red stars.

Spirals, on the other hand, developed more slowly. They formed from clouds of gas that were rotating. Because of their rotation they eventually became disk-shaped. The gas in the outer regions rotated rapidly, making it difficult for it to condense into stars. Star formation was therefore delayed in these regions, and, in fact, it is still going on today. The case for the core, however, was different. The gas was dense and the rotation was slow here so that stars in the core developed at about the same rate as they did in ellipticals.

Most astronomers agree on the main features of this scenario. It's the details that pose a problem. There are, in fact, so many problems that a number of astronomers have looked for alternatives. Among the first to do so were Alar and Juri Toomre. Once they were convinced that tidal effects were important in collisions, they began to wonder about the effects of a merger.

"It was obvious to us that there might be more to collisions than just a flyby," said Alar. "Maybe a merger occurs. At least it seemed reasonable to us to wonder about it. We weren't too cocksure it would happen, but we decided there was a good case for it." With this in mind, the Toomres made the bold assumption that ellipticals were formed in the collision and merger of two disk (spiral) galaxies. But almost no one took their suggestion seriously.

"I sat back and laughed, watching everyone sit around and say 'This is silly . . . this is stupid.' And now, little by little, astronomers are beginning to think maybe we were right," said Alar. "My own feeling is that the tide hasn't turned yet, but it's come a long way."

ARE ELLIPTICALS CREATED IN MERGERS?

One of the first to take Toomre's suggestion seriously was François Schweizer of the Carnegie Institution of Washington. Born in Bern, Switzerland, Schweizer became interested in astronomy at an early age. When he was 12 he found a booklet in his father's library titled "Astronomy on the Bernese Wavelength." It was based on a series of radio talks that had been given by a local schoolteacher. Schweizer was fascinated by it; he read it again and again. "I just loved that book," he said. He told his father he would like to meet the author and, by chance, a few weeks later, while he and his father were hiking in the Alps, they ran into him. "He was a tremendous help . . . he helped me with star maps, and he showed me how I could join the Bernese Amateur Astronomical Society," said Schweizer.

Upon graduation from high school, Schweizer went to the University of Bern to study astronomy. His first project, which involved the use of celestial mechanics, was a study of the gaps in the distribution of the asteroids. This helped him get a one-year fellowship to study at Berkeley in the United States. But when he got to Berkeley he was so enchanted with the campus and the atmosphere that he said to himself, "I'm going to get my Ph.D. here." And, indeed, he did. In fact, it was here that he first came across Toomre's paper on colliding galaxies.

"I was working on spiral density waves in 1972 when my professor, Ivan King, got a preprint from Toomre. He said to

François Schweizer.

me, 'Hey, this might be of interest to you. Why don't you take a look at it.' So I did, and I was instantly captivated," said Schweizer. "I had always hoped to understand the evolution of galaxies. And this was the first time I saw a mechanism that showed that galaxies could change their shape . . . could form tails and spiral arms. It fascinated me."

When Schweizer graduated, he took a two-year fellowship at Mount Wilson. He then accepted a position as resident astronomer at the Cerro Tololo Inter-American Observatory in Chile. By now he had become convinced that ellipticals were indeed formed by mergers of disk galaxies. But he had to find enough evidence to prove it. And in the southern skies he found a peculiar galaxy that he was sure he could build a case around. It is known as NGC 7252. It was irregular, had tails and ripples; in fact, it looked so much like the "Atoms for Peace" emblem (a bird surrounded by an atom) that it became known as the Atoms for Peace galaxy.

There was no doubt in Schweizer's mind that this was the

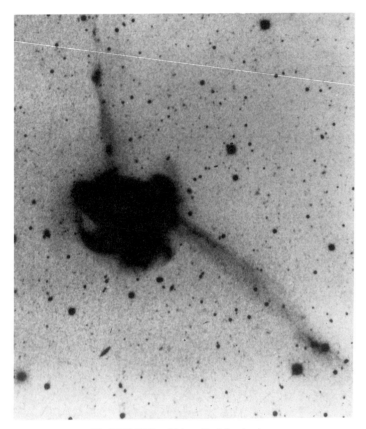

The NGC 7252 collision. (F. Schweizer)

last stage of two merging galaxies. He began studying it in detail: photographing it with various filters, taking spectroscopic measurements, and so on. Finally he found what he was looking for. At the center was a small well-defined disk of rotation, but further out the material rotated in an entirely different direction. And, even stranger, in the outermost regions the rotation was again similar to that of the core. This was virtually impossible for a single object (an object that formed from a single gas cloud). It had to be the remnant of two merged galaxies. Two long tails jutting from the system allowed Schweizer to determine when the collision occurred. Also, the fact that there were two tails told him that it had to be a collision of two galaxies of roughly equal mass. He determined that they were spirals about the size of the Milky Way that began merging about one billion years ago. They had now coalesced into a single, rather strange-looking mass. Schweizer feels that this is the first stage of an elliptical; he believes that in a few billion years it will look like the ellipticals we now see in the sky.

I asked Schweizer about it. "It's widely accepted now that it is indeed a merger remnant," he said. But he admitted that not everyone agrees that it is the first stage of an elliptical. He went on to tell me about an important discovery related to it that had just been made. "Three independent groups have now discovered that in about 30% of ellipticals the core rotates differently from the rest of the galaxy." He said that he was surprised that no one had suggested that they were caused by a major merger. "Most thought that a little galaxy had fallen into a larger one," he said. "I showed that several of them resembled NGC 7252. Then, by analogy, I argued that it is a mistake to think you can drop a small rotating galaxy into a larger galaxy and it will continue rotating. Numerical work has shown that the small galaxy will get destroyed. I think the counter-rotating cores are the signatures of newly built gas disks that are the result of two nearly equal mass spirals colliding with one another."

But do mergers create ellipticals? Or, taking it a step further: Are all ellipticals a result of mergers? Schweizer answered this with an emphatic "Yes . . . I believe they are. Ten years ago everybody doubted this. They didn't think ellipticals had anything to do with mergers. But at the Heidelberg conference in May [1989] at least half of those present thought that this was the case. It hasn't been proved beyond a doubt yet, but it has become widely accepted."

In an effort to get a more general opinion I asked several

other astronomers who are working in the area what they thought.

"Yes, I'm convinced that all ellipticals were formed by mergers," said Paul Schechter of MIT. Interestingly, he expressed surprise when I told him there was considerable controversy.

William Keel of the University of Alabama, who has been working on colliding galaxies for about six years, said, "More and more people are starting to believe it. One of the reasons is: If you look at the number of galaxies that are interacting now, you find that there are only a few. But if you assume any reasonable initial conditions [for the early universe] there would have been many such systems in the past. So where are they all now? Elliptical galaxies are about the only reasonable candidates."

Kirk Borne of the Space Telescope Science Institute in Baltimore is less certain. "I think there is a case for some of the ellipticals," he said. "There are people who believe that all ellipticals came from mergers, but I'm more cautious about this . . . although I don't think it's impossible. Elliptical galaxies are quite rounded, and they contain little gas, so they were probably formed in some kind of merger . . . something that caused a lot of gas to be used up in making stars. But the question is: When did it happen? If it happened in the very early universe nobody would worry, since everything was gaseous then, anyway. On the other hand, if you go to the other extreme and say that they were formed recently, that's when I start having my doubts."

OPPOSING VIEWS

James Gunn of Princeton University is more emphatic. "I don't think it's right," he said. "It's a controversy that has been going on for quite some time."

Born in Texas in 1938, Gunn's early life was unsettled. His father worked for an oil company and was continually on the move. Nevertheless, most of what Jim learned during this time,

James Gunn.

he said, was from his father. Together they built model trains, tunnels, landscaping, and even entire model towns.

Gunn first became interested in astronomy at age seven when his father gave him a book on astronomy. Soon they were building a telescope together. Gunn was fascinated by what he saw when they finished it: the moons of Jupiter, craters on the moon, myriad stars along the Milky Way. But when he was in the fifth grade his life was shattered. He came home from school one day to find an ambulance parked in the driveway. His father had died.

But the enthusiasm for astronomy that his father had instilled in him did not die. Throughout high school he continued to build telescopes. And when he graduated he went to Rice University where he built more sophisticated telescopes. He now equipped them with cameras and electronic equipment, and his photographs were soon so good that they were featured in *Sky and Telescope*.

He graduated at the top of his class in math and physics, then went to Caltech for graduate work. At Caltech he discovered relativity, cosmology, and the big bang theory. And soon he was hooked. Over the years he has worked in many areas of astronomy, including merging and colliding galaxies.

Gunn explained to me why he is skeptical that mergers created ellipticals—particularly, all ellipticals. "The basic issue is whether you can take objects like disk galaxies, which are not very dense, and create dense, tightly bound systems like ellipticals from them. Several people, including myself, have stressed that this is a very difficult thing to do dynamically because of what is called the conservation of phase density. This principle says that if you have objects in which the density is low, you can't increase a quantity called phase density. In the case of spirals and ellipticals, this is just the density of matter. So you can't increase the density in a collision that conserves energy. Of course, if the galaxy is made mostly of gas, energy doesn't have to be conserved. Gas can radiate. And in the early universe galaxies had much more gas in them than they do now. So you probably had this kind of merger then . . . and it likely made elliptical galaxies. But I still don't think that galaxies like present-day spirals can merge today and make galaxies like the ellipticals we see today."

Gunn is, in fact, presently working on the formation of galaxies in the early universe. Using complex computer programs, he and his students have been studying the merging of gas clouds. "We're getting objects about the size of galaxies that form stars, spirals . . . the whole works," Gunn said. He emphasized that they were arising quite naturally, without making any extra ad hoc assuptions. Despite this, though, he said he was still unable to resolve the problem of form. "We still don't really know what makes ellipticals elliptical and spirals spiral."

Jeremiah Ostriker of Princeton University shares Gunn's opinion. Born and raised in New York City, Ostriker did his undergraduate work at Harvard, and obtained his Ph.D. from the University of Chicago. He was one of the first to point out that cannibalism (the "eating" of one galaxy by another) was important in galactic clusters. "There's no question in my mind that mergers do occur," he said. "I would say that a few percent

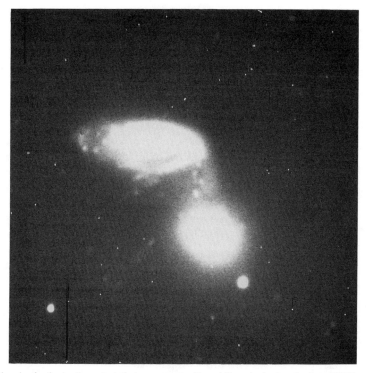

A pair of galaxies in contact that may merge. One of the galaxies is a Seyfert. (William Keel)

of galaxies undergo major mergers, and that a small amount of accretion occurs in all galaxies. But there are strong arguments that say that there can't have been a significant amount of merging of most large galaxies after they existed as complete systems. If you want to merge gas balls [in the early universe], that's quite a different story. That's everybody's picture of galaxy formation. But if you say they were galaxies like the ones we now see . . . there are many things wrong with that."

Ostriker went on to describe some of the problems to me.

Low-mass galaxies tend to be blue and have low metal content, he said, whereas high-mass galaxies tend to be red and have high metal content. "You can't add up a number of blue blocks and get a red one. If you look at the systematic properties of elliptical galaxies, they're not what you would expect from mergers."

Sidney van den Bergh of the Dominion Observatory in Canada has, in fact, shown that there are observational problems with the suggestion. One of the problems he pointed out is related to globular clusters. All galaxies have globular clusters around them, but ellipticals have many more than spirals. Ellipticals that were created in mergers would therefore have considerably less than observed.

Another problem relates to the density of the two objects. "Ellipticals are much denser in the center than spirals," said Ostriker. "You can show that when you add galaxies together you can't increase density. So you could never get the high densities you see in ellipticals."

Simon White of the University of Arizona says, "I don't think it has been demonstrated at all. I think some ellipticals must have resulted from mergers because we see the mergers and there isn't anything else they could turn into. But I think it's far from proven for all elliptical galaxies."

White pointed out what he believes is one of the major problems. "One of the difficulties is defining what you mean by a merger," he said. "The cloud that made the galaxy was probably making stars as it collapsed, and in the final stages it was likely pretty messy. It would, in fact, look like an irregular system with a lot of lumps. You can obviously look at this as a very inhomogenous collapse, or as the merger of a series of fragments."

NUCLEAR BULGES

Schweizer is well aware of the criticism. But it has not deterred him. In fact, he believes his argument can be taken a step further. If you look at the centers of most spiral galaxies you see a bulge. Indeed, as I mentioned earlier, this bulge is virtually indistinguishable from an elliptical galaxy. It is much more spherical than the overall galaxy, it contains mostly old stars, and it has no gas. Schweizer feels that these nuclear bulges were also likely formed in collisions.

"Probably all galaxies came together as bits and pieces," he said. "None of the early gas clouds were large enough to make an entire galaxy. In the history of the growth process, though, you had a few gas clouds that had no major mergers. They formed disks with no bulges . . . such as M33. On the other hand, you also had cases such as the Milky Way; it has a significant bulge." He paused briefly. "Suppose, for example," he said, "you had the Magellanic Clouds, with about 4% of the mass of the Milky Way, fall into it. You wouldn't have enough mass to make an elliptical. It would probably just fall to the center, stir things up a bit, and make a bulge."

Schweizer went on to point out two pieces of evidence in support of this idea. "First of all, about 5% of disk galaxies have blue bulges. Their spectra tell us that they consist mostly of young stars. But if the bulges were really all formed early in the universe as the old picture wants us to believe, how can you have a brand new bulge—less than a billion years old?"

Schweizer acknowledged that there are arguments against this. "I know," he said, "some people say there are just a few of them, so they're not important. I've looked into the literature and I'm convinced that there are, indeed, only a few as young as one billion years, but there are a considerable number two to three billion years old. And by five billion years, well . . . age determinations are difficult. We can't distinguish a five-billion-year-old galaxy from a 15-billion-year-old one."

Schweizer's second piece of evidence has to do with the image of the bulges. It has been known for some time that not all nuclear bulges are elliptical in shape. Some of the ones that are seen edge-on are slightly squarish. Schweizer refers to it as "boxiness." Recently, in fact, several groups have shown that up to 20% of bulges may be boxy.

So the question is: How do you form such a structure? P. J. Quinn of Australia has suggested that it may be due to a smaller galaxy falling into a large disk galaxy. He has shown that if the smaller companion falls in at an angle of, say, 45 degrees, it will precess as a spinning top precesses (i.e., its spin axis traces out a cone). And when you superimpose this on the material that is already there, you will get a "boxy" bulge.

Schweizer mentioned that "boxiness" has also been discovered in ellipticals and he believes that it likely occurs for the same reason.

But what if Toomre's suggestion that all ellipticals are created in mergers is correct? What would the consequences be? One of the most important consequences is that it means that some elliptical galaxies are relatively young. Yet, in traditional theories they are considered to be the old, or at least the same age as spirals. So it's obviously important to find out if the suggestion is correct.

FURTHER EVIDENCE: RIPPLES

Although there is still considerable controversy as to whether mergers create ellipticals, there is considerable evidence that mergers do, indeed, occur in ellipticals. An interesting line of evidence came in the early 1980s when D. F. Malin of Australia discovered mysterious-looking ripples in a large number of southern galaxies. Schweizer had, in fact, already suggested that ripples may be the signatures of mergers. Malin and D. Carter published a paper in 1983 pointing out 137 galaxies that exhibited ripples. They showed that the ripples were associated almost entirely with ellipticals. They also showed that ripples occurred mainly in galaxies that were not in clusters and, furthermore, that they do not completely encircle the galaxy (they appear as concentric arcs on either side). P. J. Quinn followed up on this suggestion as part of his thesis at the National University of Canberra. Using computer simulations, he showed

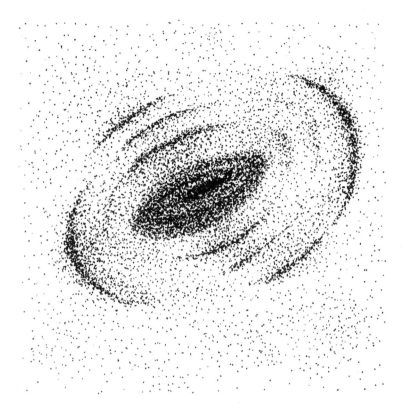

A simple representation of "ripples" in a galaxy.

that ripples were likely formed in the merger of a small disk galaxy with a larger elliptical, where the smaller galaxy spiraled into the larger one, leaving a "wake" of matter in its path.

Soon after hearing about the discovery of ripples, Schweizer began a study in collaboration with Patrick Seitzer of Kitt Peak National Observatory. It took them five years and has just recently been completed. Schweizer and Seitzer concluded that the number of ripples increased with the luminosity of the galaxy, the number of anomalies (defects) it contained, and the number of young stars in it.

Schweizer is now convinced that ripples are common in ellipticals. "If you have a really good telescope and detector, I think you would find that at least half of ellipticals have them," he said. "If you then put this together with the fact that ripples live only one or two billion years, it tells you that, on the average, an elliptical must have swallowed many smaller galaxies over the age of the universe. This seems to indicate that ellipticals are made of many small chunks." He paused briefly. "I agreed with this at first, but I'm less certain of it now. I'm suspicious that what we are actually seeing may be the result of one large merger, whose aftereffects last four or five billion years, as opposed to four or five smaller mergers that last only one billion years."

Astronomers still don't know which of these cases is true. Nevertheless it is safe to say that they now have considerable evidence that ripples are, indeed, created in mergers.

POLAR-RING GALAXIES

Another line of evidence for collisions and possible mergers comes from what are called polar-ring galaxies. The first to be discovered was NGC 2685. It appeared to be an S0 galaxy (a transition galaxy between spirals and ellipticals) with a barely visible, wispy ring around it. The main component looked like a cigar. But was it really cigar-shaped, or were we just seeing a pancake edge-on? Astronomers had not yet positively identified a cigar-shaped elliptical, and this seemed to be a good candidate.

In 1975 Marie-Helene Ulrich of the University of Texas took a series of measurements along the minor axis. They were difficult to make, but she announced that it appeared as if the rotation was around the major axis. In other words, the galaxy was, indeed, cigar-shaped. But she cautioned that further work was needed.

A simple representation of a polar-ring galaxy.

James Gunn and Paul Schecter were interested in the motion of ellipticals at the time. "We were at Palomar doing a project studying the dynamics of elliptical galaxies shortly after we heard about Ulrich's work," said Schecter. "During the course of the evening the Milky Way was overhead, and NGC 2685 was accessible so we thought we would look at it." They were fascinated by what they saw, and soon began a series of spectral shift measurements on it. They found that the central region rotated, not like a cigar, but like a pancake seen edge-on. Furthermore, the ring rotated perpendicularly to the main body.

Astronomers soon found several other polar-ring galaxies. In each case the rotation of the ring was perpendicular to the rotation of the main body. In other words, they had two planes of rotation—one at right angles to the other. But how was this possible? According to Schweizer, there is only one answer: The objects had to have undergone a collision. The two galaxies could have passed close to one another, robbing one another of gas and stars. Some of this material could have ended up forming a ring. Or an entire galaxy could have been captured—a small one. This is, in fact, the only reasonable explanation when the polar-ring galaxy is isolated in the sky.

Of course, we still have to explain why the ring is in the polar direction. To do this, let's consider the material collapsing into the galaxy. In the case of polar-ring galaxies, some of it would have had to come in over the poles. It can be shown that all of the material, except that at or near the poles, would soon be pulled into the equator. This is what happened in the case of Saturn. Its ring was likely caused by the explosion of a moon that got too close to the surface. The original material was no doubt scattered considerably, but in time it was forced into a thin ring at the equator.

Several groups have shown that material at 20 degrees from the poles would take two to three billion years to settle into the equator. And for very small angles it would take a time equal to the age of the universe. So it's easy to understand why we see rings only in this direction. You might wonder, though, if most of the material goes into a ring at the equator, why don't we also see a ring here? I asked Schweizer about this. "You do have equatorial rings," he said. He went on to say that he had discovered one a few years ago. "Recently I was remeasuring this ring when I discovered something strange," he said. "I found the galaxy rotated one way, and the equatorial ring rotated the opposite way. Five or six systems of this type have now been discovered." He hesitated, then added: "How could nature get things rotating in opposite directions in the same plane out of one gas cloud? It has to be a merger."

Another problem associated with polar-ring galaxies is: Why do they all seem to be associated with S0 galaxies? Jim Gunn offered an explanation of this. "If the center galaxy was a spiral," he said, "the gas of the disk would extend out far enough so that the gas in the polar ring would collide with it and destroy it. It seems that a gas-free S0 or elliptical galaxy is needed if the ring is to last a long time."

Gunn also pointed out that these systems have been useful in that they have allowed astronomers, through a study of the rotation of the ring, to determine how much dark matter exists around the galaxy.

OBSERVATIONS OF INTERACTIONS IN PROGRESS

Another astronomer who has been interested in interacting and colliding galaxies for many years is Kirk Borne, now of the Space Telescope Science Institute in Baltimore. Born in Baton Rouge, Louisiana, Borne was part of an air force family that moved frequently. When he was in the fourth grade he received an astronomy book as a gift from an uncle. He read it, and soon was excited about astronomy. In high school he joined an astronomy club and learned a considerable amount about astronomy. By then he was convinced that he wanted to become an astronomer. So, after completing a bachelor's degree in physics at Louisiana State University he headed for Caltech. "When I was first at Caltech," he said, "I was mainly interested in binary stars. Then one day I attended a seminar on current problems in extragalactic astronomy. After the lecture I started looking into the subject. I read some articles on it and they fascinated me. Needless to say I quickly dropped binary stars and began working on galaxies." He chuckled. "I spent two solid weeks in the library reading everything I could on colliding galaxies. I found there was quite a lack in the field. A lot of people were observing, and discovering things, but they didn't really have any way of describing them in terms of models."

Kirk Borne.

He began working on a thesis under Jim Gunn in 1982, which involved numerical simulations of interactions, but soon found that he was not satisfied with doing only theoretical work. He wanted to find observational support for his models. Collaborating with John Hoessel he therefore began a series of observations to check on his numerical work. He continued this work when he went to the Carnegie Institution of Washington in 1983.

One of the first interactions they studied was a pair known as K99. They were particularly interested in the rotation curves of the two galaxies (a plot of how the speed of rotation of the stars and gas varies as you move outward in the galaxy). The larger of the two galaxies (NGC 1587) appeared to have a normal rotation curve, indicating that it was spinning in the usual way. But when they checked the smaller of the two galaxies they were

The system K99. (Kirk Borne)

amazed. All of the outer gas and stars were being pulled off in the direction of the larger galaxy. Its entire outer layer was, in fact, being stripped off. This was the first spectroscopic evidence for such an interaction.

Borne and Hoessel then turned to another system, known as K564. In this case the two galaxies were closer together. Spectroscopic measurements soon showed that material was being pulled off both galaxies; in each case it was moving in the direction of the other galaxy. They looked for further galaxies in which this occurred and soon found several. They are now convinced that galaxies that are close to one another in space strongly influence one another, even when there is no visual evidence of interaction.

STARBURSTS, SEYFERTS, AND QUASARS

Earlier I mentioned that when two galaxies collide there is no direct collision of the stars: nevertheless, the gas clouds within the two galaxies do collide. And, in spirals, approximately 10% to 20% of the mass in the arms is in the form of gas. So we would expect it to be important.

Let's turn, then, to the collision of the gas clouds. What would happen? Considerable radiation would be given off, and we would also get a sudden increase in the production of stars. In fact, a "starburst" (a sudden burst of large numbers of new stars) would appear near the center of the colliding galaxy.

As long ago as the mid-1950s Zwicky noticed that starbursts occurred in some of the interacting systems he was studying. Arp noticed the same thing when he was assembling his *Atlas of Peculiar Galaxies*. Furthermore, the Toomres emphasized in their 1972 paper that a sudden increase in star formation during collision would likely occur. But dealing with the collision of two gas clouds was much more difficult than dealing with interactions of stars. The calculations were so complex that no one attempted them.

Then, in 1983, IRAS (the infrared satellite) was launched. One of the first discoveries was galaxies with starbursts. In almost all cases these starbursts appeared in interacting or merging galaxies. Millions of new stars were being born near the core, each emitting ultraviolet and visible light. But the new stars were immersed in dust and gas that absorbed virtually all of the ultraviolet and visible radiation. We see only the infrared radiation that is emitted when the ultraviolet and visible radiation is absorbed.

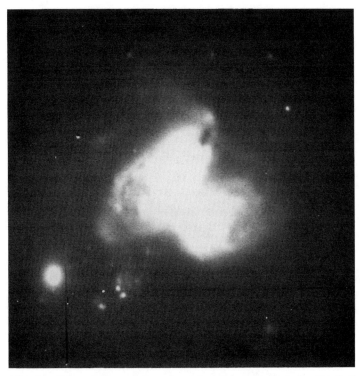

A "starburst" is occurring near the center of these two merging galaxies. (W. Keel)

But starbursts aren't the only thing that brighten the nuclei of galaxies. Seyfert galaxies, for example, have even brighter, more compact nuclei. They usually vary in brightness, and their spectrum indicates that ionized gas is moving at high speeds in the region of the core.

What makes them so intense? Again it appears as if a collision plays some role, but the intensity is so great it can't be caused merely by the interaction of two gas clouds. Most astronomers believe that a black hole is involved in this case. As we have seen, active galaxies appear to have massive black holes at

William Keel.

their centers. In fact, we saw earlier that even inactive galaxies
may have a black hole at their center. In this case the black hole
just sits there; it has "eaten" all the stars in its vicinity and no
longer has any fuel. But when a collision occurs it is refueled
and therefore suddenly springs to life with renewed brightness
at the core. Many astronomers believe that this is what happens
in the case of a Seyfert.

To check on the idea, William Keel, who is now at the
University of Alabama, and several colleagues undertook a statistical analysis of galaxies in pairs in 1982. Born and raised in
Nashville, Keel obtained his bachelor's degree from Vanderbilt.
He says he has been interested in astronomy "as far back as I can
remember." Family members tell me that my grandfather held
me up on the porch when I was about three, and pointed out the
rising moon. I think that's when my interest began."

Keel obtained his Ph.D. from the University of California at
Santa Cruz. Shortly after completing his thesis he got together
with Rob Kennicutt of the University of Minnesota, Thijs van
der Hulst of Westerbork Radio Observatory in Holland, and Ko
Hummel of the Max Planck Institute for Radio Astronomy in

Germany, and began a statistical study of interacting galaxies. "It was a shock when we discovered that a proper statistical study hadn't been done," Keel said. "We began by defining several samples of galaxies so we could perform some tests."

They selected several groups for study: One consisted of pairs of galaxies (moderately interacting, or with no sign of interaction), one was from Arp's Atlas (strongly interacting). And a third control group of isolated galaxies was selected to be used as a standard of comparison. They looked at optical spectra, radio maps of the nuclei, photometric images, and infrared data from IRAS. And their results were quite startling. "We found that galaxies that were in pairs . . . even if they didn't look distorted . . . had more Seyfert nuclei than you would expect if they were isolated. Also, they had higher star formation rates," said Keel. In the case of the Arp galaxies they found the nuclear activity was even greater. Up to 25% of the systems showed Seyfert nuclei.

They concluded that both starbusts and Seyfert activity were considerably enhanced by collisions or interactions. It appeared to them as if the collision was refueling the black hole at the center of the galaxy, causing a surge of new activity.

But if this is, indeed, true, we should be able to take it one step further. As we saw earlier, there are also indications that quasar activity is triggered by collisions. I asked Keel if he thought quasars were fueled in the same way. "Since quasars are just high energy versions of Seyfert galaxies, we would expect them to be the same," he said.

John Hutchings of the Dominion Astrophysical Observatory in Canada, and a number of collaborators, have, in fact, made an extensive study of the question. Using one of the telescopes at Mauna Kea, Hawaii, they photographed 77 quasars. They found that one-third of them looked as if a collision had occurred—they were fuzzy. In some cases, in fact, tidal interactions—arms and so on—could be seen.

"It's not at all unlikely that the quasars we observe now are all objects that are being fueled by collisions," said Jim Gunn. "A collision is a mechanism by which large amounts of gas from the outside of the galaxy can be introduced into the middle of the galaxy where the black hole is."

Computer models have, in the past few years, become extremely complex and sophisticated. And you merely have to compare the computer simulation of a given collision to a photograph of it to see how successful they have become.

CHAPTER 11

Collisions and Computers

In an earlier chapter we saw that Alar and Juri Toomre were able to show that tidal forces could draw out long tails and bridges. Furthermore, they got good matches to several systems that appeared to be interacting, including the Antennae. But they weren't the only ones working on the problem in the early 1970s. Alan Wright of York University in Toronto, Canada, was also doing a similar project. He had seen the photographs of interacting galaxies in Vorontsov-Velyaminov's Atlas and had become intrigued with them. But he disagreed with Vorontsov-Velyaminov's interpretation of how they had come about. He felt so strongly about it he set out to try and prove him wrong. So, just as the Toomres had done, he set up models of collisions on the computer. His disks consisted of 160 points representing the stars of the galaxy. Furthermore, like the Toomres, he considered all the mass to be concentrated at the center of the galaxy.

He soon found that tails and bridges would, indeed, be produced in a collision. In fact, the two galaxies didn't have to actually strike one another. "The results of these computations suggest that at least some of the deductions drawn by Vorontsov-Velyaminov are incorrect." Wright wrote. He went on to say that Vorontsov-Velyaminov was also wrong in his assumption that tails and bridges were formed in the same way spiral arms were.

It was a few years before anyone followed up on Wright's

(and the Toomres') work. In the late 1970s, however, Simon White, who was at Cambridge University at the time, became interested in the problem. He was talking to Alar Toomre and Donald Lynden-Bell in the library one day when the discussion turned to colliding galaxies. "We started arguing around and around about how mergers would occur and what the final result would look like," White said. "Alar was sure that it would look like an elliptical galaxy, and Donald was saying, 'How do you know for sure?' It soon became clear that the only way to find out was to do some numerical calculations. I was just finishing up a project I had been doing . . . a study of the evolution of galaxy clusters. So I thought I would look into the problem."

Toomre and Wright's simulations had not dealt with all the interactions that would actually occur in a collision of two galaxies. As two galaxies approach one another, all of the stars feel both the gravitational pull of the other stars in its galaxy, and the stars in the approaching galaxy. And each of these interactions have to be taken into account as the two galaxies collide.

But White realized that another problem existed. For several years evidence had been accumulating that spiral galaxies were surrounded by a large halo of dark matter. It appeared, in fact, that there was up to 10 times as much dark matter in the halo as there was in the visible part of the galaxy. This halo would obviously have to be taken into account.

Rather than using disk-shaped galaxies, as Wright and the Toomres had, White began with spherical objects that represented the sphere of dark matter. White admits, though, that one of the reasons for selecting spheres was that they were incapable of doing much more at the time. "We had no idea what would happen so we started with the simplest case," said White. He used 250 points in each galaxy, and soon found that the halo would indeed have a serious effect.

Within a short time, a number of other astronomers began similar simulations. Soon galaxies were represented, not just by a simple sphere, but by a disk with a spherical halo around it to represent the halo of dark matter.

One of White's later projects was a simulation similar to

Toomre's of the Antennae. White included the effect of the halo, but he was not satisfied with the results. "It was just too crude," he said. "It was qualitatively correct, but we didn't get a picture that would impress anyone. We didn't get nice tails . . . just fat ones that weren't long like the real ones."

In an effort to test Toomre's suggestion that elliptical galaxies are created in mergers, White also did a series of experiments with colliding galaxies designed to see if a collision would leave its "signature"—something in the appearance of the disk that could be recognized later as resulting from a collision. He concluded that it would not.

JOSHUA BARNES

In the early 1980s White took on a graduate student by the name of Joshua Barnes. Barnes had become interested in galaxies, and the collisions of galaxies many years earlier while an undergraduate at Harvard. "The first thing I saw on colliding galaxies was the Toomres' article in *Scientific American*," said Barnes. "I looked at it and said to myself, 'I'll bet I could do that.' And, in fact, I did . . . it wasn't too hard at all."

At the time Barnes was interested in both computer science and astronomy. He was fascinated by computers, but didn't want to become a computer scientist because of his interest in astronomy. So he spent a lot of time in the computer center while taking basic physics courses.

But he was impatient to get into research. "After a couple of years I switched into astrophysics," he said. "In physics there is so much you have to learn to get started in research. But in astrophysics there are lots of problems and few people to do them, so that even an undergraduate can do something. So I made the switch, and although I managed to do some interesting research [as an undergraduate] none of it got published." The breakthroughs that he made later on, however, he feels were definitely an outgrowth of what he did at this time.

Upon graduation, Barnes went to graduate school at the

Joshua Barnes.

University of California at Santa Cruz. Despite his fascination with computers and astronomy he decided to work in relativity. "That was a mistake," he said. "There just aren't enough good problems to work on in relativity. There are a few things . . . but I just didn't see where I would fit in." So after a couple of years he decided to chuck it and go to Cambridge, England. At Cambridge he began working on galaxies. This was more to his liking; he felt much more at home with galaxies than he did with relativity. Besides, he was using computers again; his project involved the dynamics of groups of galaxies. Within a year he had made considerable progress.

Upon returning to the United States he went to Berkeley where he met Simon White. White was also interested in computer modeling of galaxy interactions, so Barnes began working on a thesis under him. White agreed that he should continue working on the project he had begun at Cambridge—the evolution of groups of galaxies. And within a couple of years he had his Ph.D.

From Berkeley Barnes went to Princeton's Institute for Advanced Study. Although most of his work to this point had been on the dynamics of groups of galaxies he had some experience with the problem of colliding galaxies. Indeed, he had duplicated some of Toomre's results many years earlier. Furthermore, White and some of his students had been working on the problem. Barnes soon became interested in seeing whether he could improve on what had already been done.

By the mid-1980s computers had become larger and faster than they had been when Alar and Juri Toomre did their work. But something more was needed. If the calculations were to give a realistic result, the number of particles used to represent the stars had to be significantly increased. Trying to represent billions of stars by a disk of a few hundred points was not realistic. You needed at least a few thousand, and preferably many more. But a large amount of computer time was required if a large number of particles were used. Indeed, if you were to include all of the interactions between all of the points (stars) of two colliding galaxy models, the number of calculations would be enormous. Computing time increases as the square of the number of particles. This means that if you want to increase the number of particles by a factor of 10, from, say, 100 to 1000, the computing time would go up by a factor of 100. So, if 100 particles took one hour to do, 1000 particles would take 100 hours.

Barnes realized that something else was needed—new techniques, or new "algorithms," as scientists refer to them. Working with Piet Hut, he began a search. They noticed that the most sensitive interactions were those between particles that were relatively close to one another. In the case of two approaching galaxies they would be the forces between the particles on the approaching sides of the disks. They therefore partitioned the particles into groups, or cells, of various sizes. When the particles were close to one another and interactions were strong, each of the individual interactions resulting from the particles within the group was taken into consideration. But when the particles were far from one another, and the interactions were weak, a group of particles was considered as a unit. As the distance between the particles increased, the group became larger.

"This technique turned out to speed things up by an important factor," said Barnes. "If you wanted to increase the number of particles from 1000 to, say, 10,000, you only had to do a little more than 10 times the amount of computing, as opposed to 100 times as much. So it really made sense to use a large number of particles."

When the new algorithm was polished and working, Barnes and Hut began thinking about how they wanted to apply the technique. But they differed on how to proceed. "Piet wanted a big research program involving lots of people, so he started out trying to generate funding for it. I didn't want to wait . . . so I simply sat down and started doing what I could. And, sure enough, the money didn't come, but the computer models did." Barnes was pleased with the results. "At Berkeley," he said, "I could only run 2000 particles. With the new algorithm I could run up to 64,000."

Barnes began with the Antennae. Alar and Juri Toomre and Simon White had modeled it earlier. Barnes was eager to find out what his new technique would give. With the large number of particles at his disposal he could now represent the major components of a spiral galaxy: the arms, the bulge at the center and the massive halo of dark matter that surrounded it. When the results finally came Barnes was pleased. There was an almost exact match to the Antennae. The tidal tails were exceedingly narrow, as they are in the photographs. And the shape of the main body matched much more closely than earlier models. Barnes then followed the merger into the future, and found that the system eventually evolved into something that looked very much like an elliptical galaxy.

Encouraged by the success of the Antennae, Barnes went on to look in detail at the effect of the dark halos around galaxies on collisions and mergers. He compared collisions with and without halos. Over 16,000 particles were used in these simulations.

"The effect of the dark matter halos was really significant," said Barnes. "In fact, if we didn't know that such halos existed, we could have discovered them with this kind of calculation. What I discovered was that the dark matter acts as a brake on the motions of the visible galaxies. The two galaxies plunge through one another and begin to separate. It's at this point that the dark matter really kicks in. It slows them down so that they end up trapped in a very tight orbit around one another. Then they fall back and have a closer passage, and they keep doing this until they merge."

The dark-matter halos were also important in that they absorbed the rotation of the two merging galaxies. "One of the major objections to merging as a way of making ellipticals has always been that the merger remnants would spin much faster than real ellipticals," said Barnes. "The dark matter resolves this problem. The two galaxies 'grab' one another as they pass, and this slows their rotation. Then after several passes the galaxies have a nearly head-on collision, and a final merger occurs. By this time there may be little rotation left."

Barnes explained that even without the dark matter the collision would give something that looked like an elliptical galaxy. They would have the right shape and brightness profile, but the motion of the stars within the object would not be correct. "The dark matter is critical in providing the correct stellar dynamics," he said. "With the kind of calculations I've been doing I've been able to produce structures that are very similar to what we see in real ellipticals: slow rotation, proper distribution of stars among the various orbits and so on."

Barnes seemed so confident that his merger simulations were, indeed, producing ellipticals, that I asked him if he was convinced that all ellipticals were created in mergers. "Yes, I think they are," he said. "If you subject a stellar system to vio-

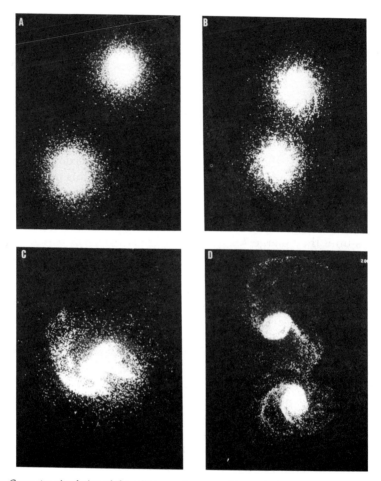

Computer simulation of the collision and merger of two galaxies. a) The approach. b) Approximately 150 million years later. c) The first encounter at 250 million years. d) Galaxies separate. Tidal forces create long filaments. Time: Approximately 375 million years.

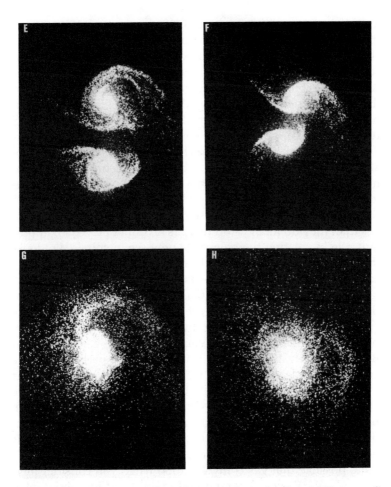

e) Galaxies circle around one another and move together again. Time: 1 billion years. f) Galaxies begin to merge again. Time: 1.25 billion years. g) Final merger begins. Time: 1.375 billion years. h) Merger remnant—an elliptical galaxy. Time: 2 billion years. (Joshua Barnes) lent shaking and churning, you tend to set up a density profile [the change in density that occurs across the galaxy] that is characteristic of an elliptical galaxy. So they had to result from some type of merger. But a merger of what? There are obviously several possibilities. It could be that all star formation occurs in disks and these disks merge together to form ellipticals. But you have to be suspicious of this right away because of the IRAS results. You've got things there that are not disk galaxies . . . and they're making stars like crazy. So that idea is probably out." He paused briefly. "The sort of picture I would put my money on is one in which a great deal of the material involved in mergers is in disklike structures. It's probable, though, that at early times these structures contained a lot more gas than they do today. And this gas no doubt played an important role in creating the bulges we now see in spirals. I'm looking forward to the day when I can make things like this in my computer, and see what happens when they collide."

Earlier I mentioned that Gunn and Ostriker were against the idea of ellipticals being created in mergers because "phase space density" could only be lowered in a merger. Yet it appeared to be much higher in ellipticals than in spirals. I asked Barnes if he felt there was any way around this problem. "There are two possible ways around it," he said. "First off, if you start with spiral galaxies that have a bulge, the bulge is like a little elliptical. So you're starting with something that has a high phase space density . . . and it's not surprising that you get it at the end." He chuckled. "I'll have to admit, though, that this begs the real question: Where did the bulges come from? It's quite possible that gas is involved. So when you consider mergers of real galaxies, it's important to think not just about what the stars will do, but also what the gas in the galaxy will do. This, unfortunately, is a hard problem . . . gas dynamics is tricky stuff."

Barnes mentioned that he is currently working with Lars Hernquist of the Institute for Advanced Study on this problem. They have found that when a collision of two galaxies that contain considerable gas occurs, a lot of the gas gets funneled to the center. As the collision begins (the first encounter) a bar is formed at the center that acts to break up the gas flow. As the gas flows around the bar the gravitational pull on it is stronger in some places than it is in others, and as a result it is not always pulled toward the center. This causes "shocks" in the gas and as a result it begins to fall inward. "We've found that you can funnel as much as 60% of all the gas into the center region in this way," said Barnes. "And it's one way of explaining why ellipticals have high phase space density. The gas is funneled to the center, and some of it turns into stars, which, in turn, increases the phase density."

Another of the arguments against ellipticals as merger remnants is that ellipticals have far more globular clusters around them than spirals do. I asked Barnes about this. "It's a very interesting puzzle," he said. "Of course there are only a few ellipticals close enough that you can count their globular clusters. So the argument isn't laden with good statistics."

Barnes pointed out that the major problem was with ellipticals that are in clusters. He felt that the problem was not serious if they were not in a cluster. He went on to say that if they were in clusters, the mergers likely occurred well in the past when the galaxies contained much more gas than they do now. And in the merger some of this gas would likely go into the making of globulars. Most of the more recent mergers, he felt, were taking place in low density regions. "It's in these low density regions that present day spirals are being scrambled up to produce ellipticals . . . and they have about the right number of globular clusters for 'field' ellipticals [not in clusters]," he said.

THE COLLISION OF M51 AND NGC 5195

Let's turn now to some of the collision simulations that Barnes and others have done on the computer. The first one we will look at involves one of the most beautiful spirals in the sky—the

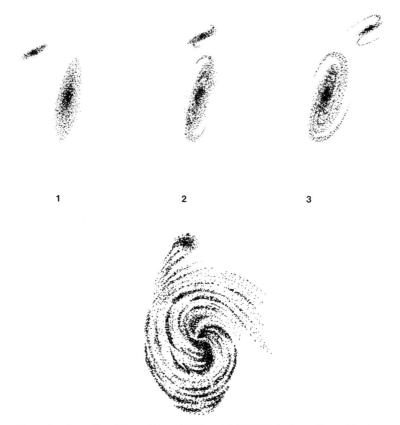

1   2   3

Computer simulation of the collision of M51 and NGC 5195. 1, 2, and 3 are side views.
Last diagram is face-on.

Whirlpool galaxy (M51). It lies in the constellation Canes Ven-
atici.

The Whirlpool has distinct, medium-loosely wound arms
that exhibit many young stars. Off to one side is a smaller galaxy
(NGC 5195), also a spiral, although considerably less regular
than M51. An arm extends from M51 to the smaller galaxy,

M51 and NGC 5195. (Lick Observatory, University of California, Santa Cruz, Calif. 95064)

indicating that there is, indeed, an interaction between them. Alar and Juri Toomre were the first to try to simulate the interaction. Beginning with two spirals they found that they were able to produce a system that looked very much like the one we see in the sky. They assumed that NGC 5195 passed by M51 several million years ago with the outer regions of the two galaxies overlapping in the flyby. To get a final result that looked like the real system they found they had to assume that the two galaxies were rotating in opposite directions.

As NGC 5195 passed by, part of one of the spiral arms of M51 was pulled after it. Stars and gas on the opposite side of M51 were also seriously affected, causing the arm on that side to move outward. In the process many stars were knocked entirely out of the galaxies, and are now flying off in intergalactic space. The shattered disk of NGC 5195 continued on past M51 and is now at some distance from it. To us, however, it appears as if it is still nearby—superimposed on it, in fact. This is because of the way we see it.

Barnes and Hernquist recently turned their attention to this encounter. Alar and Juri Toomre had got the general outline of the collision, but Barnes wanted to reproduce it much more closely. The Toomres, for example, hadn't been able to duplicate the inner spiral structure. Also, recently radio telescopes have shown an enormous spiral arm (made up mostly of hydrogen gas) that extends out from the main body of M51 well beyond NGC 5195. "We have had a fair degree of success in matching this arm," said Barnes. "But as of yet we don't have the inner spirals in exactly the right place. But they're close . . . so it's encouraging. It's hard to match exactly. You have to start with exactly the right amount of mass in the two models, and you have to have the right profile. All you can do is guess and follow it through to see what happens. And if it doesn't work you make changes and do it again."

THE COLLISION OF NGC 7252

In the last chapter I talked about some of the early work of Kirk Borne. He has continued to work in the field and recently has done an interesting simulation of the collision and merger of NGC 7252. I asked him how many particles he was now using in

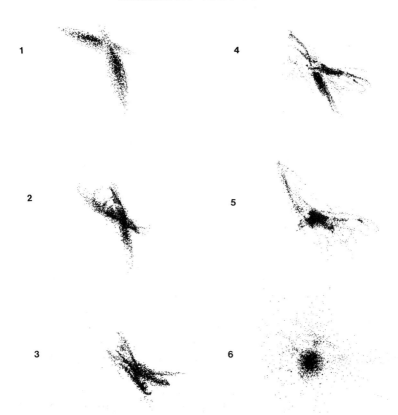

Computer simulation of the collision of NGC 7252. Compare number 5 with figure on page 196.

his models. "We usually only use a few hundred when we're doing a test," he said. "But in the final analysis we use several thousand. I've gone up to about 10,000," His work on NGC 7252 was done in conjunction with Douglas Richstone of the University of Michigan. "Schweizer provided the observations we used in our model," said Borne. "He had some very strange observations. We tried to model them."

As in the case of M51-NGC 5195, the Toomres were also the

first to try to model NGC 7252. But little was known about the system at that time.

Borne and Richstone represented each of the galaxies by a disk of 2000 stars. In the actual collision the galaxies had to have had roughly the same mass. Alar and Juri Toomre showed that long tails such as those seen in NGC 7252 would occur only in the collision of galaxies of roughly the same mass. If one of the galaxies had been much less massive than the other, only one tail would have been visible.

Borne and his colleague found that the best results—the closest match to observations—came from disks that were rotating opposite to the direction of their orbital revolution, yet they had to be at a considerable angle to one another.

Although only one collision was simulated by Borne and Richstone, it was clear from the work of Barnes that more than one has actually occurred. Barnes showed that galaxies usually undergo several collisions before they actually merge. Two galaxies that approach one another spinning in opposite directions will, in fact, separate again soon after the initial contact. They will pull back and separate by about the diameter of one of the galaxies. The gravitational pull between them will, however, soon overcome the outward inertia and they will be pulled into one another again. It is this final collision and merger that Borne and Richstone are primarily concerned with. They found that it was relatively easy with their models to match the two long narrow tails that are seen in the photographs of the object. In fact, they even managed to match the observed speeds of the tails with their model.

The major difficulty was the body of the merged galaxies. Schweizer had shown that there were two regions that rotated differently. With a considerable amount of work, however, they were able to duplicate it.

One of the major results of the simulation was the projection of the merger into the future. Borne and Richstone found that in about a billion years it would become symmetric and would closely resemble an elliptical.

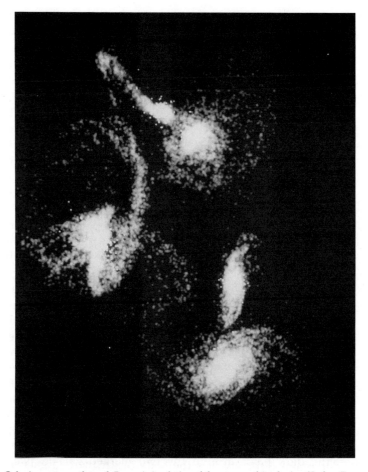

Galaxies partway through Barnes' simulation of the merger of 6 galaxies. (Joshua Barnes)

THE COLLISION AND MERGER OF SEVERAL GALAXIES

Much of Barnes' early work with galaxies involved the dynamics of groups of galaxies. So it is perhaps natural that he eventually turned back to this problem. And, indeed, after completing his

work on the effects of dark matter and central bulges on collisions and mergers of two galaxies, Barnes turned his attention
to the mutual collision of several galaxies.

Scattered throughout space are a large number of small
groups or clusters of galaxies that are bound to one another.
Each of the galaxies within these groups has a particular speed
in a particular direction, and it is highly probable that it will
eventually hit one of the other members of the group. Barnes
wondered what would eventually happen to such a system. He
therefore selected six disk galaxies and computed their evolution over several billion years.

"I didn't start off with a group where the galaxies were
falling toward one another," said Barnes. "I started with an
equilibrium situation, so that the galaxies were given random
orbits. But they were gravitationally bound as a unit. Then, as
they interacted with one another, their orbits decayed. The dark
matter played an important role in this. It acted to break and
slow the galaxies down, and that brought them closer together,
until finally they all merged."

As in his previous work each of the galaxies were taken to
consist of a disk, bulge, and halo. He put roughly four times as
many particles in the halo as in the other two regions. In his six
systems there was a total of 65,536 particles.

Barnes' group of six consisted of four that were the same
size, and two that were twice as big as the others. He found that
after approximately 400 million years two of the galaxies merged
with two of the others, leaving only four behind. During this
time two of the four also developed long narrow tails. Furthermore, the dark halos around the two merged galaxies were
much larger after the merger than they were before.

At 1.2 billion years the two merged galaxies underwent a
close encounter that severely disturbed their halos. The largest
remnant then went to the center of the group. The four remain-
ing galaxies continued moving until at 2.4 billion years one of
the smaller ones merged with the large one at the center. This
left only one small disk galaxy orbiting the huge one at the

center. It was soon stripped of much of its mass. Finally at 3.6 billion years the small one fell into the giant, and the only thing left was a huge elliptical.

THE MILKY WAY–MAGELLANIC CLOUDS INTERACTION

If there is considerable evidence that many of the galaxies around us are interacting we might wonder if perhaps the Milky Way is involved in some sort of interaction. And indeed there is evidence that it is. The evidence is indirect and far from conclusive; nevertheless, many astronomers are now convinced that the Milky Way is interacting with our nearest neighbor in intergalactic space—the Magellanic Clouds.

The first hint that something strange was going on came in the 1960s when astronomers discovered that many of the gas clouds in the outer sections of our galaxy were far above the disk. Furthermore, they had exceedingly high velocities. In short, it appeared as if the disk of the Milky Way was "warped."

Then, in 1973, Don Mathewson and several colleagues in

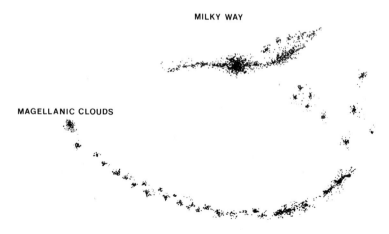

The Magellanic Stream.

Australia discovered a long stream of gas in the southern sky. It stretched across the entire sky and ended at the Magellanic Clouds. Astronomers were soon referring to it as the Magellanic Stream. It curved back beneath the Milky Way and was about 300,000 light-years long, but there was no evidence that it was connected to the Milky Way. Then astronomers found a group of clouds near the end of the stream. They had exceedingly high velocities and seemed to be patches of gas that had been torn from the end of the main stream. If this was, indeed, the case, the Magellanic Stream looped all the way from the Magellanic Clouds to the Milky Way. The two systems were, indeed, connected. Furthermore, it seemed as if the high-speed clouds in the outer regions of the Milky Way were also connected with this loop.

But what would have caused this? The most likely explanation is that the stream is a tail that was ripped out of the Magellanic Clouds in a near miss of the two galaxies that occurred about 200 million years ago. If this is indeed the case the two systems may eventually start to move back toward one another again in another encounter. They may, in fact, eventually merge.

CHAPTER 12

Clusters and Superclusters

So far we've considered turmoil within individual galaxies, in other words, exploding galaxies, and in the collision of galaxies. But there is also considerable turmoil on a much larger scale. After all, galaxies aren't the largest things in the universe. Clusters of galaxies are larger, and we now know that clusters of clusters exist. Furthermore, we shouldn't forget the most turbulent event of all—the big bang. Let's begin by considering clusters of galaxies.

One of the first plots of large scale structure, interestingly, was done before astronomers realized that galaxies existed. In 1921 C.V.L. Charlier of Sweden plotted up 11,475 objects from J. Dreyer's New General Catalog and noticed that there was considerable clumping of the points. A few years later Hubble showed that most of these objects were galaxies, and astronomers began to wonder if galaxies perhaps came in groups, or clusters.

In the early 1930s Hubble decided to look into this. He began by photographing 1283 sample regions of the sky with the 60 and 100-inch reflectors at Mount Wilson. He then counted the number of galaxies of each magnitude per unit area of the sky and compared them in various directions. In all, he counted 44,000. He found that there were no galaxies in the direction of the Milky Way. Furthermore, there were few close to its edge. But in a direction perpendicular to it there were large numbers.

His first conclusion was that the gas and dust of our galaxy were blocking our view of the galaxies behind it. Indeed, even near its edge many galaxies were obscured. Making allowances for this obscuration he concluded that the number of galaxies was roughly the same in all directions. He then counted the number of galaxies as far out into space as his telescopes could penetrate and concluded that they were homogeneously distributed.

But when he looked at the distribution statistically, he found that it was not the case. The slight "clumping" that was noticeable was statistically relevant. Many, if not most, of the galaxies were in clusters.

Fritz Zwicky came to the same conclusion about the same time. He noticed that there was a considerable range in the size of clusters. Some contained only a few members while others contained hundreds.

THE LOCAL GROUP

It soon became obvious that our galaxy, the Milky Way, was also part of a cluster of galaxies. We now refer to it as the Local Group. It has about 30 members, most of them dwarf galaxies.

We are lucky in that we are able to see stars in several of the other members. And some of them are cepheid variables. This allows us to use the cepheid period-luminosity relation (we talked about it in Chapter 2) to determine their distance. Also of help are the globular clusters that surround several of the members. Walter Baade of Mount Wilson Observatory showed in the 1940s that the brightest stars in globular clusters are all about the same brightness. Since we can see some of these stars, we can determine the distance to the galaxy they are associated with.

The nearest members of our group are the Magellanic Clouds, two irregular galaxies that are about 200,000 light-years away. We saw in the last chapter that they appear to be interacting with the Milky Way. They are also interacting with one another; a stream of gas can be seen in the region between them. Stars, in fact, have recently been detected within this stream.

The Andromeda galaxy. (National Optical Astronomy Observatories, Tucson, Arizona)

The largest member of our group is the Andromeda galaxy (M31), which lies at a distance of 2 million light-years. It is a spiral similar to the Milky Way that is slightly inclined to our line of sight. It consists of a bright nucleus surrounded by long spiral arms that contain many young stars and considerable gas and dust. Nearby is another spiral (M33). Its arms are much more loosely wound.

The Milky Way is the second largest member of the Local Group. It lies at one end of the group, the Andromeda galaxy lies at the other end. As the two giants of the group, they control the motions of all the other members. Both are surrounded by a group of dwarf galaxies. In all, there is about two dozen of these small galaxies. One of them is the dwarf irregular known as Barnard's galaxy, named for the astronomer who discovered it, Edward Barnard. (It was the subject of Hubble's first paper on galaxies.) Star clusters and gaseous nebulae can easily be seen in it. About the same distance away is the dwarf elliptical NGC 6822 which is similar in many ways to Barnard's galaxy, except that it is smaller.

The Andromeda galaxy has two conspicuous companions: M32 and NGC 205. M32, an elliptical, is interesting in that it appears to be interacting with M31. The spiral arm closest to M32 is disturbed. Computer models have, in fact, shown that the distortions in M31 could have been caused by a small galaxy such as M32. There are also indications that M31 has stripped away many of the stars from its smaller neighbor.

M31's other conspicuous companion, a dwarf elliptical, also seems to be interacting with it. Ellipticals are normally symmetric and contain little, if any, gas or dust. NGC 205 is not symmetric—it has a distinct twist along its length—and it has considerable gas and dust in which new stars are forming. The twist appears to be caused by tidal forces from M31. But where did the gas and dust come from? A number of astronomers believe that they may have been pulled from M31.

Two other members of the Local Group, Sculptor and Fornax, were discovered by Shapley and his assistants in the late 1930s. Sculptor was noticed first. It was so faint and unusual that Shapley didn't know what it was. He thought it might just be a stellar group within our galaxy. But his assistants showed that some of its stars were cepheids. In the meantime, another similar object was found in the constellation Fornax that also

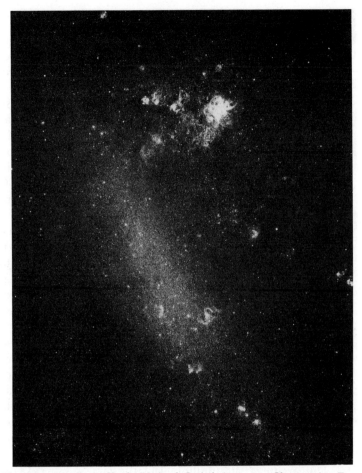

The Large Magellanic Cloud. (National Optical Astronomy Observatories, Tucson, Arizona)

had visible cepheids. Both were soon shown to be dwarf ellipticals that are members of the Local Group.

Although many astronomers have studied the Local Group, one of the most extensive studies has been done by Paul Hodge

A small cluster of galaxies. (Hale Observatories)

of the University of Washington. He has determined the distances of several of the members and also made a detailed study of their properties. Hodge first became interested in astronomy while in high school in Snohomish, Washington. "I went to the library to see what they had in astronomy," he said. "There was nothing that had been published since 1914, so I took the bus to Seattle and went to the University of Washington bookstore where I found Shapley's book *Galaxies*. I spent all my allowance on it. It had a wonderful chapter on the Local Group . . . and I've been interested in it ever since."

Upon graduation from high school Hodge went to Yale where be obtained a bachelor's degree. He then went on to Harvard where he did a thesis on the Magellanic Clouds under Cecilia Payne-Gaposchkin. After obtaining his Ph.D. he went to Mount Wilson for a year, then taught at Berkeley briefly. Finally, in 1965 he and a colleague went to the University of Washington where they set up the department of astronomy.

At the present time Hodge is studying the rate of star formation in several of the members of the Local Group in an effort to determine how it has varied over the lifetime of the cluster. "We hope to find out if star formation in galaxies of different types has been continuous, or if it is highly sporadic, or perhaps decreasing." he said. To answer this question he is making measurements on the youngest stars in some of the nearby galaxies, then looking at increasingly older groups to see how the rate of star formation differs.

I asked Hodge what was the latest count on the number of members of the Local Group. "It rises and falls, a little like the stock market," he said with a laugh. "That's because the distance to a few of the objects is still uncertain. I'd say there are 25 for sure, and about 5 or 6 we're not absolutely sure about." The major problem, he said, is getting reliable distances for some of the small galaxies.

OTHER CLUSTERS

Our group is, of course, only one of many. Two of the larger nearby groups were identified as early as 1908 by the Swedish astronomer C.V.L. Charlier. He noticed that there was a considerable amount of clumping in the constellations Virgo and Coma Berenices. Today we realize that both of these clumps are clusters; in fact, both are much larger than our Local Group. The

Virgo cluster, an irregular group, is the largest in our region of space. It contains about 1500 galaxies and lies about 40,000 light-years away in the direction of the constellation Virgo.

The Virgo cluster contains both spiral and elliptical galaxies, and has three giant ellipticals (M84, M86, M87) near its center. Each contain trillions of stars. The only objects that can definitely be identified as spirals lie in the outer regions. A close examination of the group, however, shows that many of the galaxies in the region between the outer edge and the center resemble spirals; they are disk-shaped, but have little gas and dust, and no arms. In fact, even at the center we find disklike objects with no gas and dust, or arms, that resemble spirals. In short, there appears to be a gradual transition from normal spirals in the outer regions to armless spirals with no gas or dust at the center.

What is the cause of this strange distribution? Certainly it has nothing to do with the normal evolution of spirals. Gas is used up continuously in the production of stars in spirals, but most spirals, including our own, still have an ample supply.

We obviously have to begin by asking ourselves if these disk galaxies are, indeed, related to spirals. And the answer seems to be yes. They appear to be spirals that have somehow lost their arms, and most of their gas and dust. Virgo, in fact, is not the only cluster that has this strange distribution. Many others also have it.

How did these clusters get this distribution? Astronomers are now confident that they can explain what has happened. X-ray observations show that Virgo, and other similar clusters, have hot (about 100 million degrees) gas throughout them. The density is highest at the center and drops off gradually as you move outward. In the outermost regions its density is relatively low.

Consider what would happen to a galaxy as it moved through this gas at a velocity of several million miles per hour. It would obviously experience considerable wind. The gas within a galaxy is, of course, held to it by gravity, but is it held strongly

Part of the Coma cluster of galaxies. (Lick Observatory, University of California, Santa Cruz, Calif. 95064)

enough? Calculations show that it isn't. Any galaxy crossing the inner one-third of the cluster would, in fact, be quickly stripped of all its gas.

On the basis of this we can now explain the variety of structures within Virgo. Because the density of intergalactic gas is low in the outer regions, the wind would be negligible, so spirals here would be able to hold onto their gas. Galaxies closer to the center, however, would experience considerable wind and would soon lose much of their gas. And finally, galaxies at the center would be scoured of all gas in a few hundred million years. Most galaxies have been around for about 15 billion years, so a hundred million years is a small fraction of their overall life.

Turning now to the Coma cluster, we find it is quite different from the Virgo cluster. It is about 500 million light-years away, generally spherical in shape, and composed entirely of ellipticals and disk galaxies. It contains over a thousand large galaxies and perhaps as many as 10,000 smaller ones. Furthermore, the density of galaxies increases significantly as you approach the center.

The Coma cluster is, in many ways, like the central region of the Virgo cluster. In other words, it contains ellipticals and a few disklike objects that look like spirals with no arms. Needless to say, the gas has likely been stripped from these galaxies in the same way it was in the case of the Virgo cluster.

Another well-known cluster lies about 700 million light-years away in the direction of the constellation Hercules. It is similar to the Virgo cluster in that it is irregular and contains both ellipticals and spirals. If you continue looking around at other clusters you will find that they are either symmetric like the Coma cluster, or irregular like the Virgo and Hercules clusters. George Abell was one of the first to notice this. While still a graduate student, he began a survey of clusters of galaxies. In all, he cataloged 2712, finding that they fell into two main classes; he called them regular and irregular.

CANNIBALISM

Something else that is frequently noticeable when you look at a photograph of clusters is huge ellipticals at the center (see pho-

Note the large ellipticals at the center of this cluster. (National Optical Astronomy Observatories, Tucson, Arizona)

tograph). As I mentioned earlier, there are three in the center of the Virgo cluster. Several are also visible at the center of the Coma cluster. They are so much larger than any of the other galaxies it makes you wonder what's going on. Furthermore, in most cases they are radio galaxies, and some of them have double or multiple nuclei.

What causes these large galaxies? As you might expect, cannibalism is no doubt involved. To put it simply: They have grown big and fat by feeding on their neighbors. And it's easy to see how this could happen. In the last chapter I talked about Barnes' computer simulation of a small cluster of galaxies. We saw there that they all eventually came together to create a huge elliptical.

Let's look at the details. The galaxies of a cluster orbit the center of mass of the cluster in much the same way that the planets of the solar system orbit the sun. Furthermore, their orbit depends on their energy. And since energy depends on both speed and mass, the more massive a galaxy, for a given energy, the slower its speed. This means that the most massive galaxies will have orbits closest to the center of mass, and when small galaxies pass them tidal forces will slow them down. They will lose energy and spiral inward to an orbit closer to the center.

Eventually, the largest galaxy will settle at the center and the others will begin to spiral closer and closer to it. As they spiral around it some of them will merge. They are now primarily under the influence of the giant at the center. As small galaxies pass close to the central galaxy, their outer regions will be stripped off, and most of the stars and gas will fall into it. In fact, large clumps of stars may fall in. According to some astronomers this is what causes the double and multiple nuclei that are sometimes seen.

In some cases, 50 or even up to 100 galaxies may have been cannibalized by the central elliptical. But I also mentioned that most of these ellipticals are strong radio sources. What causes this? As you likely have guessed, most astronomers believe a huge black hole at the center of the elliptical is responsible. When a small galaxy falls into this elliptical it feeds the black hole and in the process creates radio emission.

Gerard de Vaucouleurs.

GERARD DE VAUCOULEURS

If clusters of galaxies are so common, we immediately wonder: Are there larger structures, such as clusters of clusters? Indeed, there are. When George Abell was assembling his atlas in 1959, he noticed that many of the clusters appeared to be grouped in clusters of clusters, or superclusters, as they are now called. But it was the French astronomer Gerard de Vaucouleurs who was the first to really push the idea. After studying many of the nearby clusters he concluded that our Local Group was part of what he called the Local Supercluster. He published the suggestion in 1953, but few astronomers paid any attention to it.

De Vaucouleurs' interest in astronomy began early. After reading Paul Couderc's book *Architecture of the Universe* in 1930, he became so fascinated with astronomy that he began reading everything he could about it. Within a short time he had a small telescope and had joined the French Astronomical Society. By 1932 he had decided to become an astronomer.

At college he took courses in astronomy, physics, and mathematics. Soon after he graduated in 1939 he met a wealthy amateur astronomer by the name of Julien Peridier who had built an observatory in the south of France. De Vaucouleurs worked for him until the outbreak of World War II. He was then called into the army, where he served until 1941. When France was occupied by the Germans he went back to his studies and began working for Peridier again in his spare time. He received his Ph.D. from the Sorbonne in 1949.

But even with a Ph.D., de Vaucouleurs didn't find things easy. He now had a wife and badly needed a job, but opportunities for astronomers in 1949 were limited. So he went to England where he got a job producing a weekly science program for the BBC. He found the work interesting; it allowed him to travel extensively and he was given free rein in programming. Still, he longed to get back into real astronomy—research. So when a job became available at the Stromlo Observatory in Australia, he took it.

In Australia he studied galaxies and clusters. He was updating a catalog of galaxies that had been completed many years earlier by Shapley and Ames when he noticed that the Milky Way was situated on the edge of a large group of clusters. He named the group the Local Supercluster. But most astronomers considered his evidence to be insufficient. Others said the idea was outright nonsense.

Although de Vaucouleurs spent most of his time in Australia studying galaxies, he also became fascinated with the planet Mars, and soon became an expert on it. About this time, Lowell Observatory in the United States began looking for an expert on Mars. De Vaucouleurs was offered the position and he accepted it. Interestingly, when he got to Lowell he spent most of his time on galaxies. He continued pushing his ideas about superclusters, but was able to generate little interest.

From Lowell he went to Harvard and for the next two years most of his efforts were directed toward Mars. He even published a book on the planet titled *Physics of the Planet Mars.* Finally, in 1960, he went to the University of Texas where he had access to the 82-inch reflector of the McDonald Observatory. At that time it was the third largest in the world. Furthermore, the demands on the telescope were minimal, so he had considerable use of it. And he made good use of his time. Along with his wife, he compiled a catalog giving information on 2600 galaxies which he published in 1964. It was a continuation of the work he had begun in Australia.

As a result of this work he became even more convinced that our Local Group was part of a supercluster. But most astronomers remained skeptical. By the time he published a second catalog containing information on 4300 more galaxies a few years later, however, the tide had begun to turn. Astronomers were finally beginning to realize he was right.

THE LOCAL SUPERCLUSTER

Most of the bright galaxies of the northern hemisphere belong to the local supercluster. The Local Group is near one edge. The center is dominated by the huge Virgo cluster we talked about earlier; it contains about 20% of the total galaxies in the system, and is about 60 million light-years from us.

Brent Tully and Richard Fisher of the University of Hawaii have mapped the Local Supercluster in considerable detail. They measured the redshifts of over 2200 galaxies and produced a three-dimensional map of the structure—a project that took them 9 years. They found two main clouds with numerous cigar-shaped clouds emerging from them. In all there are millions of galaxies in the system. It is about 100 million light-years

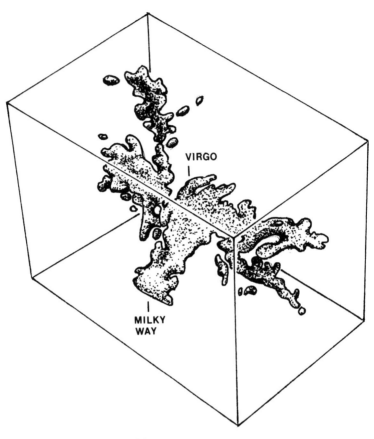

The Local Supercluster.

long by 10 million light-years thick. Strangely, though, it is mostly empty space; 98% of the overall mass is in 11 clouds.

OTHER SUPERCLUSTERS

What about other superclusters? Are there any? It might seem, with ours being so large, that there would be little room left over for others. But this isn't the case. In 1976 Stephen Gregory, now of the University of New Mexico, and William Tifft of the University of Arizona, were studying the Coma cluster when they discovered it was part of a supercluster.

Born and raised in central Illinois, Stephen Gregory became interested in science at the age of 9. Einstein's death had a great impact on him. "I heard a lot about him at the time," he said. "I was impressed . . . I wanted to be like him." Then, a few years later, Sputnik was launched, and his enthusiasm for space and the universe increased. Upon graduation from high school he went to the University of Illinois where he got his bachelor's degree. He then went to the University of Arizona for graduate work.

He wanted to become a theorist in astrophysics. "As a thesis project I decided to construct a theoretical model of the Coma cluster, so I went to the library to see what was known about it. I wanted to have a target . . . some observations to point my theory toward. But I found out there was almost nothing known about it. It immediately hit me that I shouldn't be doing theory, I should be doing observations. And I've stuck with observational work ever since."

He therefore undertook an observational study of the Coma cluster. He began by taking redshift measurements of all the component galaxies. His object at this stage was to determine which galaxies were members of the cluster and which weren't, and to find out something about the dynamics of the system. "When I looked over my initial data," he said, "I found out it was important to look at galaxies that were in the foreground because they seemed to have a odd distribution. We expected this region to be filled with galaxies. Instead, to our surprise, there was a huge void. My advisor, Bill Tifft, was the first to see the odd distribution in my data."

At first he couldn't believe it. But after rechecking he found it was indeed true: There was an immense region of space that contained literally no galaxies. It was strange, but Gregory assumed it was just an anomaly.

Gregory continued his study of the Coma cluster and, in 1978, along with Laird Thompson, who is now at the University of Illinois, he found a chain of galaxies that appeared to be connected to it. Following the chain they came to another large cluster called A1367. Checking the redshift of A1367 they found it was the same as that of the Coma cluster; this meant that the two clusters were the same distance away. The galaxies between them did, indeed, form a bridge between them. Studying the system further, they found it was a chainlike structure that stretched across space for a distance of 100 million light-years. Although dominated by two huge clusters, it was indeed a supercluster.

In 1981 Gregory, along with Laird Thompson and William Tifft of the University of Arizona, discovered another large supercluster in Perseus. It was about 200 million light-years away. It spans about 40 degrees in the sky, part of which lies in the constellation Pisces. This system may be joined to another system in Lynx and Ursa Major that was discovered by Martha Hayes and Riccardo Giovanelli of Cornell, using radio telescopes. It too is at a distance of about 200 million light-years, and it seems to line up with it. We cannot see the region where they would join because it is cut off from our view by the Milky Way. But if they do, in fact, join, the overall system would be immense.

There is also another large system in Hercules. It is about 500 million light-years from us and, like the Coma supercluster, it appears to have a large empty region in front of it. To Gregory and his colleagues, it was, in fact, the voids that were the big surprise. "The discovery of the superclusters was not unexpected," said Gregory. "In fact, we were pretty sure we would find one. But the discovery of the voids . . . they were unexpected."

Then, in the late 1970s, Robert Kirchner of the University of Michigan, Augustus Oemler of Yale, and Paul Schecter of Mount Wilson Observatory were studying a region of the sky in the direction of the constellation Böotes. They wanted to determine the average number of galaxies per unit volume in this region of the sky. As they moved outward to dimmer and dimmer galaxies they came upon a region that contained virtually no galaxies. At first they were annoyed because it would cause a problem in their average. But when they looked into it further, they found it to be incredibly large: a spherical region about 300 million light-years across. This was considerably larger than the void found by Gregory and his colleagues.

We now know that there are many such voids in the universe. And between the voids are superclusters. I asked Gregory how many superclusters there were. He hesitated, then said, "That's a number that can't really be stated with any confidence. They're all over the place . . . but there may be only one huge one. They may all be interconnected. We don't know for sure."

The real problem, however, is how these superclusters and voids were formed. We will look at this in the next chapter.

CHAPTER 13

To the Ends of the Universe

We turn finally to the universe as a whole. Turbulence, as we will see, also plays an important role here. We begin by asking: What does the universe on the grandest scale look like? In other words, if we could somehow step back from it and take a look, what would we see? In the last few years, astronomers have been able to do this, and they have found a fascinating structure. You would, of course, expect voids and superclusters. But they have found much more. The whole universe seems to be filled with voids; in fact, they are more like gigantic bubbles. And strung along the surface of these bubbles are superclusters. It sounds crazy, but on this scale, the universe looks like a sink full of soapsuds.

How did this structure form? Obviously there had to be considerable turbulence. But what would cause this turbulence? Astronomers are still not certain. They have learned a lot about the large-scale structure of the universe in the last few years, and many interesting theories have been put forward to explain it. But there is still much to be learned, and the problem is complex. Let's begin by looking at some of the difficulties.

LOOKBACK TIME

When we look into the universe we don't see it as it is today. Why? Simply because we are looking at images that are composed of light, and this light takes a finite time to reach us. For us to see all parts of the universe as they are now, light would have to travel at an infinite speed, and it doesn't; it travels at 186,000 miles/sec. This means that if we look at a star that is 10 light-years away, we see it as it was 10 years ago. For distant stars, this time delay can be large, but it pales in comparison to the delay that occurs in the case of galaxies. The Andromeda galaxy, for example, is two million light-years away; its light therefore takes two million years to get to us. This means that if you looked at it tonight you would see it as it was two million years ago. And the Andromeda galaxy is one of the nearest. There are galaxies hundreds of millions and even billions of light-years away.

But why is this a problem? The major reason is because galaxies evolve in time. If you looked at a distant galaxy through a telescope tonight you would see it as it looked hundreds of millions of years ago. If you could see it as it really appeared today it would no doubt look quite different. When galaxies are young they are filled with young stars, blue giants, supernovae, and so on. They were therefore much brighter then than they are today.

But how do we know that distant galaxies are actually younger? If the universe were infinitely old, this wouldn't be the case. But we don't believe the universe is infinitely old. According to the big bang theory, the theory that we presently accept, the universe was "born" about 18 billion years ago, and the first galaxies appeared in it a few hundred million years later. Furthermore, these galaxies have changed as they have aged. This means that if we look out into the universe (back into time) we are actually seeing its history. If we had a large enough telescope we could, in theory, look right back to the big bang.

How far, in fact, can we look back? As we saw in an earlier chapter, quasars appear to be the most distant objects in the universe. Recently we have discovered one as far away as 14 billion light-years. It is obviously very close to the big bang. It turns out, however, that we will never be able to see all the way back to the big bang. The reason: for the first hundred thousand years the universe was opaque; in other other words, it was "foggy." At this stage the universe was so dense that the radiation was coupled, or locked, to the matter. It wasn't until the temperature got down to 3000 K that this radiation was able to free itself and expand out into the universe. When this happened the universe suddenly became transparent.

What's the significance of this? It means that if we look far enough we will eventually see only the opaque cloud; the big bang is buried beyond it.

CURVATURE OF THE UNIVERSE

Something else that has an effect on what we see is the curvature of the universe. Einstein's general theory of relativity tells us that the universe is curved by the matter in it. This curvature, in fact, depends on the amount of matter it contains. At a density called the critical density (2×10^{-29} gms/cm^3) the universe is flat. If the density is greater than this, the universe is positively curved, and distant galaxies will actually be larger than they appear. If the density is less, the universe is negatively curved, and galaxies are smaller than they appear.

Curvature is also important in relation to the fate of the universe. If the universe is positively curved it will eventually stop expanding and collapse back on itself, shrinking eventually to a point. If it is negatively curved, on the other hand, it will continue expanding forever.

So far we're not sure which of these scenarios will occur. Observations indicate that the universe is negatively curved by a a relatively large factor, but there is theoretical evidence that it may be flat. One of the major problems is the "dark matter" that permeates the universe. This is matter we are sure is there, but can't identify. It contributes to the overall mass; in fact, 90% of the mass of the universe may be in the form of dark matter. And until we resolve the problems associated with it we cannot say for certain how the universe is curved.

We see, therefore, that many things have to be taken into

consideration if we are to understand the overall structure of the universe, but in many ways this makes the problem more fascinating. Let's turn, then, to this structure.

THE LARGE-SCALE STRUCTURE

One of the first large-scale plots of galaxies was done by Jim Peebles and several of his students at Princeton University in the early 1970s. Peebles used data that had been accumulated by C. Donald Shane and Carl Wirtanen of Lick Observatory. Over a period of several years they photographed the entire northern hemisphere.

Peebles plotted the galaxies from the photographic plates. Actually, he never plotted individual galaxies; he used a shade of gray at each point depending on how many galaxies were at that point. He used dark gray if there were many, light gray if few. In all, Peebles and his students plotted approximately a million galaxies and what they got was breathtaking. The universe had an amazing structure: Filaments or chains were visible along with knots and regions that contained few galaxies. It was mottled, and looked like something you would get if you threw mud at the side of a white house.

But did the universe actually have a structure like this? It was possible, but astronomers couldn't be sure from the plot. The problem was that it was two-dimensional; this meant that the galaxies were plotted on top of one another. What we needed was a three dimensional plot so we could see how the galaxies were stretched out in space.

How could we get this? The simplest way would be to measure the redshifts of all of the galaxies, then use a Hubble plot (a plot of redshift versus distance) to obtain their distances. But in the 1970s determining redshifts was still a long and tedious task. Obtaining the redshift of a single dim galaxy could take hours.

The structure obtained by Peebles was a surprise, but within a short time other evidence was found that substantiated it.

The first voids were discovered, and then the first superclusters. But if astronomers were to find out what the overall structure really was like, a large-scale redshift survey involving thousands of redshift measurements was needed. And for this to be possible techniques had to be improved. Something more than just photographic plates was needed. By the late 1970s astronomers had, indeed, developed several new devices.

One of the first to initiate a large-scale redshift survey was John Huchra, who is now at the Harvard-Smithsonian Center for Astrophysics; he later joined forces with Marc Davis, who is now at the University of California, and Margaret Geller (she left for England for two years shortly after the group was formed). After working for several years building new detectors, spectroscopic equipment, and computers to run the system, Huchra and Davis were ready to start taking data in 1978. Within a few years they had the first three-dimensional maps of the extragalactic universe. The filamentary structure—the chains of galaxies and huge voids—seen by Peebles were still there. The universe was, indeed, mottled, even in three dimensions.

Davis left the group in the late 1970s and went off to the University of California. But Huchra was determined to continue the work, so he teamed up with Margaret Geller who had returned from England. They were assisted by graduate student Valerie de Lapparent. In an effort to penetrate farther into the universe they began taking data in "slices." These slices resemble a piece of pie; they are 6 degrees thick by 135 degrees wide. The velocity derived from the redshift is plotted in the radial direction (outward from the center of the pie).

In 1986 they completed their first slice and made their first discovery. The universe in three dimensions was mottled, but it was mottled in a strange way. It looked like soap bubbles. The bubbles, or voids, were spherical, roughly 150 million light-years across, and forming their surfaces were huge strings of superclusters.

Huchra and Geller continued making measurements in slice after slice. Then came a second discovery.

Margaret Geller and John Huchra.

THE GREAT WALL

By 1989 Huchra and Geller had measured more than 10,000 redshifts, completing four slices and partially finishing several others. What they found was truly amazing. Cutting across each of the slices was a huge feature. It was a gigantic structure—a thin sheet of galaxies 500 million light-years long, 200 million light-years wide, and approximately 15 million light-years thick. It was about 200–300 million light-years from the Earth. Several astronomers, including Marc Davis of Berkeley and Avashai Ockel of Israel, dubbed it the "Great Wall."

It's not visible to the naked eye, or even with a moderate-sized telescope, yet it runs from horizon to horizon across the entire northern hemisphere. It runs through, and therefore contains, the Coma supercluster. I asked Huchra to describe it to me. "It has many groups and clusters in it," he said. "We see it going across four slices. It runs into the galactic plane on the

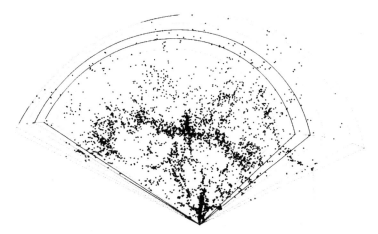

The Great Wall (dark region through the center). (Geller and Huchra)

ends, so we're not seeing it all. There's a good chance, in fact, that it goes all around the sky. Whether it's a coherent single entity or whether it's something that shows up by chance because of the way things are distributed is hard to say. Anyway, it's awfully big."

He went on to say that it appeared to be strung out across the surface of bubbles. One thing that surprised him was how thin it was. "The best way of thinking about it is as a quilt," he said. It has the same dimensions—big in two dimensions, and extremely thin in the other."

THE GREAT ATTRACTOR

The Great Wall, it turns out, may not be the only "wall" of galaxies in the universe. There is evidence for another one— called the "Great Attractor."

Let's consider the evidence. As we have seen, the big bang theory tells us that the universe is expanding: that every galaxy, including our own, is moving away from every other galaxy. We can, in fact, determine the velocity of a given galaxy of known distance merely by looking at its Hubble plot (a plot of redshift versus distance). This plot is usually assumed to be a straight line. But if you look at it you see there is considerable scatter in the points, even for nearby galaxies. At first astronomers were convinced that this was due to the difficulties of obtaining accurate distances, particularly for distant galaxies. But if this is, indeed, the case, why is there also a considerable scatter for nearby galaxies?

Astronomers soon began to realize that, in addition to the expansion of space resulting from the big bang (called the Hubble flow), galaxies also had peculiar velocities due to their uneven distribution. Galaxies and clusters of galaxies are gravitationally attracted to other galaxies and clusters that lie nearby, and as a result they move toward them. These peculiar velocities, as they are called, can in fact be fairly large. And until we understand them we are going to have difficulties formulating a theory of how galaxies, clusters, and superclusters formed.

Let's take a closer look at this problem. Before the mid-1960s there were strong arguments that peculiar velocities were important, but proof was lacking. Then came the discovery of the cosmic background radiation. As we saw earlier, this is radiation that was given off when the universe cooled to a temperature of about 3000 K. As the universe expanded, this radiation cooled until it reached its current temperature of 2.7 K. When Penzias and Wilson of Bell Labs discovered it, they found it was uniform in all directions to an accuracy of about one part in 1000. In 1977, however, with improved instrumentation and techniques, astronomers discovered that it was not exactly uniform. In one direction it had a temperature of a few thousandths of a degree hotter than the average; in the opposite direction it was cooler by the same amount.

What was the significance of this? Assume for a moment that you are surrounded by fog. What would happen if you began moving through it? If you took a few measurements you would find that it appeared to be more dense in the direction you were moving. In the same way, if you were in a constant temperature bath, the temperature would be slightly greater in the direction you were traveling. This meant that the Milky Way galaxy was moving relative to the background radiation; in fact, its velocity was calculated to be 600 km/sec. Astronomers then began looking at the velocities of the other nearby galaxies in our Local Group. All of their velocities seemed to be directed towards the center of the group—the direction of the huge Virgo cluster. Galaxies on the other end of it also had peculiar velocities in its direction. Soon there was no doubt: We were "falling" into the center of our Local Group. But when a further study was made, astronomers found that this could not account for 600 km/sec; it could only account for 250 km/sec.

Astronomers then began looking for other objects that might also be attracting the Milky Way (and the Local Group). The most logical source would be another nearby supercluster. Our nearest supercluster is the Hydra-Centaurus supercluster, so astronomers immediately began a study of the region of space around us that included the Hydra-Centaurus supercluster. The result: Our Local Group was indeed moving toward the Hydra-Centaurus supercluster, but to their surprise astronomers found that the Hydra-Centaurus supercluster itself was also moving relative to the microwave background with a speed even faster than that of the Local Group. Furthermore, both superclusters were moving in the same direction.

A strange result. What did it mean? Obviously there had to be something beyond the Hydra-Centaurus supercluster that was attracting both it and the Local Group. Astronomers now refer to it as the "Great Attractor." Measurements indicate that this Great Attractor has to be at least twice as far away as the Hydra-Centaurus supercluster. If it is, indeed, at this distance its pull on our cluster indicates that it has to have tens of thousands of excess galaxies (in excess of the average distribution around us).

One of the first reactions upon hearing about such an object is: What about the Great Wall discovered by Geller and Huchra? Is it possible that it is the Great Attractor? I asked Huchra about this. "No, it can't be the Great Attractor," he said, "It's in the wrong direction. But it is possible that the Great Attractor is something like this." So far, though, he said, no deep space surveys have been made in the direction of the Hydra-Centaurus supercluster.

WHAT CAUSED THE STRUCTURE?

But if the universe is composed of huge "walls" of galaxies interspersed with bubbles, we have to ask ourselves: What caused this structure? The bubbles are spherical, so it might seem that they were generated by an explosion. On the other hand, it is quite possible that they are a result of gravity.

One of the first to tackle the problem was Yakov Zel'dovich. Interestingly, when he first became involved with it in the early 1970s little was known about the large-scale structure of the universe, although there was a vague notion that superclusters existed.

Zel'dovich became an astrophysicist literally by accident. Upon graduation from school he began working as a lab assistant at an ore-processing laboratory in Leningrad. He was visiting the Leningrad Physical-Technical Institute one day in 1931 when he began discussing a problem with several of the institute members. They were so impressed with his knowledge of metallurgy and chemistry they asked if he could be transferred to the institute permanently. The ore-processing lab was reluctant to let him go, but after some negotiation he was traded for a vacuum pump.

A few years later he received a degree equivalent to our Ph.D. Most of his work during this time was on explosives and shock waves; some of it, in fact, had important military applications. His real interest, however, was physics, and with the discovery of nuclear fission in 1939 he turned his attention to the theory behind the process. He later worked on nuclear reactors and, together with Andrei Sakharov, he was a central figure in the development of the Soviet atomic and hydrogen bombs. After the war he turned to elementary particles and astrophysics. Of particular interest to him were the discoveries being made in the large-scale structure of the universe. The explanation of this strange structure was a challenge he couldn't pass up.

It soon became obvious to him that the structure had to be related to fluctuations, or clumps, in the early universe. This meant that the cloud coming out of the big bang had to develop irregularities soon after it emerged. Slight compressions and rarefactions (regions of low density) must have developed. He realized that as the cloud expanded, matter would be pulled towards the compressions, and out of the rarefactions. This would cause a breakup of the cloud. According to Zel'dovich it broke up into clouds that eventually condensed into superclusters. He showed, in fact, that the collapse would be much more rapid along one axis than it would along the others, and as a result huge "pancakes" would form. Astronomers now refer to his theory as the pancake theory.

According to this theory, clusters and superclusters formed along the regions where the pancakes intersected. The interior of the pancakes were the voids, and the regions of intersection were the superclusters.

But there was a serious problem with Zel'dovich's theory. It took longer than the age of the universe to form the structure in it. The problem perturbed Zel'dovich, but he was sure it could be overcome. He looked for a way around it, and in 1980 he heard that the neutrino (a ghostly particle that barely interacts with matter) might have a mass. This was exactly what he needed. Incorporating it into his theory and allowing for it to stream through the universe he was able to show that neutrinos would "blow away" all fluctuations except those that were of the right size to produce superclusters. The result was a universe of voids and chains of superclusters.

But even with neutrinos, difficulties remained. We saw earlier that if clusters and superclusters are to form, fluctuations are needed in the original gas cloud. Calculations show, however, that if you put small fluctuations in the cloud, structures form rapidly—too rapidly. In fact, they soon become gigantic black holes. Gravity, it seems, works too well! If structures similar to those we see today were to form, the cloud would have had to have been extremely smooth. In fact, it could not have differed from region to region by more than one part in about 10,000. This means that it had to have been almost perfectly uniform, yet it couldn't have been exactly uniform.

How do we know it had to be so smooth? We merely have to look at the cosmic background radiation. Even now it is almost perfectly smooth, varying only by about one part in 10,000. This leads to the enigma: How could the matter of the universe clump and form galaxies, clusters and so on, yet leave the cosmic background radiation so smooth? Zel'dovich's theory doesn't explain this.

But Zel'dovich's theory isn't the only one we have. In addition to Zel'dovich's theory, which assumes that the largest structures—the superclusters—formed first, then clusters and galaxies later, we have a theory formulated by Jim Peebles that takes the opposite point of view. In this theory a roughly uniform distribution of galaxies formed first, then some of the galaxies began attracting others and clusters began to form. Finally clusters attracted other clusters and superclusters formed. Again, as in the case of Zel'dovich's theory, we end with chains of superclusters and voids. But time is again a problem. It seems as if not enough time has passed to produce this entire sequence.

Peebles' theory is usually referred to as the bottom-up theory because it starts with the smallest structures first, forming the larger ones later. Zel'dovich's theory, on the other hand, is referred to as the top-down theory. Peebles' theory is also unable to explain how galaxies could form from fluctuations and leave the cosmic background radiation so smooth. In short, both theories have serious problems, and because of this other theories have been put forward. In order to understand these new theories we must first look at the big bang in more detail.

THE BIG BANG

One of the major problems with the big bang theory is that at the very earliest stages we have no way of explaining what went on. Before a time of 10^{-43} seconds, called the Planck time, the theory that we normally use to describe the universe, namely general relativity, breaks down. At this stage the universe was so small, a quantized version of general relativity is needed to explain the events that occur. But so far we don't have one. This means that we can only speculate on what happened.

Let's turn, then, to what happened after the Planck time. I'll give a brief description of each of the important events in the order that they occurred. As the universe expanded, it of course cooled. About 1/10,000 of a second after the big bang, the temperature had dropped to a trillion degrees. At this stage the universe consisted of electrons, photons, neutrinos, and a few protons and neutrons. About 15 seconds later the temperature was down to a few billion degrees. Protons and neutrons were still present, but they soon decayed into electrons and other light particles. As the universe continued to expand, it continued to cool, and after about 3 minutes the first nuclei appeared and collisions created some of the light elements.

As we saw earlier, at this point the universe was still opaque. After about 100,000 years the temperature was down to about 3000 degrees and the nuclei began to capture electrons. When this happened the radiation broke free from the matter, expanded off into the universe, and space suddenly became transparent. We now see this radiation as the cosmic background radiation; it has a temperature of approximately 2.7 K. Finally, fluctuations developed and matter began to clump, eventually forming galaxies, clusters, and superclusters.

In a nutshell, then, that's it. It's what the big bang theory tells us, and the big bang theory is the best theory we have of the origin of the universe. But there are several things that it does not explain. One, called the flatness problem, was noticed by Robert Dicke and Jim Peebles in 1979. They found that if the universe is relatively flat today, say within a factor of 2, then one second after the big bang, its density had to be equal to the critical density to one part in about a million billion. In short, the early universe had to be incredibly flat.

INFLATION AND QUANTUM FLUCTUATIONS

One of those who became interested in the shortcomings of the big bang theory was a postdoctoral student at Cornell by the name of Alan Guth. In 1979 he began looking at the details of the early universe. He soon saw that as the universe cooled it went through several phase transitions just as water changes from steam to liquid water to ice as it cools. Upon examining the mathematics that describe these changes, Guth found that supercooling may have occurred as the universe passed through an early transition. If so, it would have created what we now call a false vacuum. From a simple point of view, this is just a vacuum that has a mass. What is particularly strange about this vacuum is that as it expands, its energy density remains constant. Guth realized that this would cause an incredibly fast expansion. The universe would double in size in about 10^{-36} seconds, then continue to double every succeeding 10^{-36} seconds.

Guth found that this inflation would last only from about 10^{-36} seconds to 10^{-34} seconds. At the end of this time the universe, which was now about the size of a basketball, would continue to expand at its usual rate. What particularly pleased him was that inflation solved not only the flatness problem but several other difficulties of the big bang theory.

As it turned out, though, inflation theory had problems of

its own. Guth found he couldn't bring the inflation to a smooth ending. Within a couple of years, however, this problem was overcome by Andrei Linde of the USSR, and independently by Andreas Albrecht and Paul Steinhardt of the United States. The revised theory is now called New Inflation.

But how could inflation have anything to do with the structure of the universe—the superclusters and voids? It soon became obvious that it would have a serious effect. First of all, any fluctuations in the universe before 10^{-36} seconds would be greatly magnified by inflation. But what type of fluctuations existed this early? How would they be generated? When astronomers began looking closely at the effects of inflation, it seemed at first as if the theory created more problems than it solved. Inflation would be so dramatic and powerful that it would smooth out any irregularities. The universe after inflation would be perfectly smooth. Furthermore, there was still the problem of where the fluctuations came from. Inflation theory didn't explain that.

Astronomers were stumped. Then in 1982 an important breakthrough was made. A conference was called at Cambridge University in England to deal with the problem of galaxy formation. During the conference someone pointed out that just before inflation the universe would be atomic-sized and tiny fluctuations, called quantum fluctuations, would occur naturally in the universe.

Before I explain why they would be important let me take a moment to explain why they occur. In the world of atoms and elementary particles, we cannot specify the position of a particle exactly (assuming its speed is known accurately) because of what is known as the Uncertainty Principle. An important consequence of this principle is that particles and antiparticles can be created out of the vacuum. They live only a short time, but during this time they create density variations—in other words, fluctuations.

Calculations were soon made and it was found that quantum fluctuations that existed before 10^{-36} seconds would be inflated rapidly. They would be smoothed out, and therefore could explain the smoothness of the cosmic background radiation, but they would still be sufficiently "clumpy" to create the structure that we now see in the universe. This was a significant breakthrough. There was now no need for assuming that fluctuations somehow mysteriously appeared in the early universe. Quantum fluctuations would be there naturally.

With the introduction of quantum fluctuations and inflation many of the problems of how galaxies, clusters, and superclusters arose finally appeared to be solved. But as the details were developed it was found that a few problems remained.

COMPUTER MODELS AND DARK MATTER

Once astronomers had solved the major problems related to the formation of galaxies, it was obvious that they would soon try to generate the large-scale structure of the universe using computers.

Within a short time, though, theorists found that a serious problem existed. Just as White, Barnes, and others found they had to include dark matter in their simulations of colliding galaxies, astronomers found that they also had to include dark matter if they hoped to match the voids and filaments that were observed.

Furthermore, it wasn't just a matter of adding an arbitrary component to represent the dark matter. The results depended critically on the form of the dark matter. And at the present time we have no idea what form it takes. Interestingly, computer simulations may be useful in this respect. We can assume a certain type of dark matter and see whether it gives filaments and voids similar to those observed. If it does we can assume that the dark matter is of this form.

In order to understand the problem better, let's briefly consider some of the dark matter candidates. One of the more important ones is the neutrino. For many years it was not considered seriously because it was assumed to have zero rest mass. But, as I mentioned earlier, in 1980 two groups announced that the neutrino might have a mass. It would be exceedingly small, less than one-thousandth the mass of the electron, but if this was the case, neutrinos would be an important dark matter candidate. So far, though, there has been considerable controversy over whether it actually has a mass, and if so, whether it is in the right range.

Other particles—axions, supersymmetric particles, and so on—have also been suggested as candidates for dark matter. They are usually referred to as "exotic particles" since none of them have been observed. They exist only on paper (i.e., they are predicted by various theories).

As far as computer models are concerned, astronomers usually divide dark matter particles into two types: hot dark matter (HDM) and cold dark matter (CDM). The neutrino is the major HDM particle, so called because of its high velocity. Axions and supersymmetric particles, on the other hand, are expected to move slowly, and are therefore CDM particles.

The first computer simulations were based on the assumption that the dark matter was HDM. Simulations of this type were made by Simon White and Marc Davis at the University of California, and independently by Adrien Mellott, who was then at the University of Pittsburgh, and Joan Centrella of the University of Illinois. Both groups worked on the project through the early 1980s, and indeed voids and chains did come out in the simulations. The results, in fact, seemed to be in fair agreement with Zel'dovich's top-down theory. But by 1983 both groups realized that HDM could not create voids and chains that matched observations.

In 1983 Davis, White, and their colleagues began to consider cold dark matter. They soon found that the agreement with observation was much better. Mellot and Centrella followed suit. Both groups found that the results seemed to agree best with Peebles' bottom-up theory.

But again the agreement was not perfect. Modifications

were made which helped. Then Davis and White introduced what they called "biased galaxy formation." According to this idea, the early universe consisted of a cloud of cold dark matter (along with a few protons and neutrons) within which were small fluctuations. As expansion continued, "density peaks," or regions of excess density were created that, in time, attracted other density peaks. The program was, in effect, biased in favor of large galaxies; galaxy formation from small clumps was suppressed.

This gave better results. But with the recent discovery of the Great Wall, even it seems to be in trouble. "The cold dark matter model is having problems," said Huchra. "If we assume the universe started out 10 to 20 billion years ago as a homogeneous cloud you can't make something as large as the Great Wall. The problem is the length of the thing . . . it's half a billion light-years long. That's 5% the diameter of the universe. You can't make something that large in the allotted time. You either have to assume the universe started off very unsmoothly, or you have to have something other than gravity being responsible for the creation of such structures. Or else you have to show the universe is actually a lot older than we think it is."

EXPLOSIVE MODELS AND COSMIC STRINGS

But if gravity can't create large structures such as the Great Wall, is there anything else that can? Jeremiah Ostriker of Princeton University and Lennox Cowie of the University of Hawaii have suggested an alternative: an explosion model. S. Ikeuchi of the University of Tokyo has also worked on a similar model. According to their idea, matter is pushed together as a result of shock waves that are generated by galaxies. When a galaxy is young, large numbers of supernovae occur. According to Ostriker and Cowie, these supernovae could generate a shock wave that would sweep out from the galaxy scouring the region around it. Such waves could, in fact, be sent out by clusters and even superclusters.

Another possibility that is closely related to this is cosmic strings. Such strings may have been generated in the early universe as it went through various phase transitions. Just as water leaves behind cracks and other imperfections when it freezes, so too would imperfections have been left behind in the early universe as it went through various phase changes. We would have regions of material from a previous phase—for example, regions of false vacuum, wedged between regions of true vacuum.

Astronomers have shown that these imperfections would take the form of massive, but extremely thin strings. These strings would either stretch across the universe or take the form of closed loops. The loops would, in fact, oscillate rapidly, creating gravitational waves that would repel the matter around them, thereby creating large voids. Better still, the oscillations would create extremely energetic electromagnetic waves that would repel the material around them. If there were a large number of strings of this type throughout space, they would obviously create large voids and push matter into filaments that would end up as chains of superclusters.

The major problem with this idea is that we have not yet detected cosmic strings.

SUMMARY

Mapping of the large-scale structure of the universe will no doubt continue for many years as astronomers push farther and farther into the universe. Theoreticians will continue trying to explain it and, as they do, new puzzles and problems will arise. But this is what we expect—it's the way science progresses.

CHAPTER 14

Epilogue

And so we come to the end of our story. In our journey through the universe we have seen an amazing array of objects: exploding galaxies (some with powerful jets), colliding and interacting galaxies with huge tidal plumes emanating from them, and quasars—distant and mysterious. Each is an awe-inspiring spectacle.

The power generated in the explosive outpouring of a galaxy is so great it is difficult to comprehend. Furthermore, in most cases the activity takes place deep in the core of a galaxy, and is therefore out of sight. But there is a way we can get some appreciation of it. We can travel to it in our imaginations. To do this, though, we have to give our imaginations free reign. The sizes are so tremendous, and the energies so incredible, that they are difficult to encompass. Nevertheless, imagination is important, particularly to the scientists who are studying these galaxies. If they are to build better models, they have to be guided by their imaginations.

Considerable progress has been made in the last few years, but there is still much to be learned. Where, we might ask, do we go from here? Have we learned most of what there is to learn about the universe? Scientists in the past have said that we were on the verge of knowing everything there is to know. But each time they were proved wrong. For the more we learn the more we realize how much more there is to learn. We may, in fact, even now, be on the threshold of a whole new series of discoveries—discoveries that will completely change our view of the universe.

Many new projects are presently being planned in astronomy; some, in fact, have already begun. The Hubble telescope, for example, has been launched. In space its 94-inch mirror is unhindered by the Earth's murky and turbulent atmosphere. With 50 times the sensitivity of the 200-inch Palomar reflector, it will give us an unprecedented view of the universe. On board will be two spectrographs, two cameras, and a photometer—all state-of-the-art.

What do we expect to learn with the Hubble telescope? Certainly our view of distant galaxies will be enhanced. Objects that now appear to be little more than tiny smudges will take on a new clarity. Furthermore, we will be able to see galaxies much farther out in space. And we will no doubt learn a great deal more about quasars, the most distant objects in the universe. As we saw earlier, most of the nearby quasars appear to be associated with colliding galaxies. Does this apply to all of them? The Hubble telescope may give us the answer.

Finally, we will get a much better view of the chaos that is occurring in the nuclei of exploding galaxies. And we will be able to study the interactions, the arms and tidal bridges, between colliding galaxies as never before.

Soon after the Hubble space telescope was launched a gamma-ray telescope called GRO will be put in orbit. It will be invaluable in studying the gamma rays emitted in high-energy processes in stars and galaxies. Black holes, quasars, and supernovae are all copious sources of gamma rays. This telescope will give us a new view of the gamma-ray universe, and many new insights into black holes, quasars, and other objects will no doubt soon follow.

Beyond GRO we are looking at the launching of the X-ray satellite AXAF (Advanced X-Ray Astrophysics Facility). It will be used to study many of the same things as GRO. Some of the most significant advances in astronomy in the last few years were made as a result of the X-ray satellite Einstein (1978), and the earlier Uhuru. AXAF will pick up where Einstein left off; it will be 100 times more sensitive, and is expected to last considerably longer.

Finally, in the infrared region, we will soon have SIRTF (Space Infrared Telescope Facility). It is a cooled telescope with a detector about 3 feet in diameter. It will be approximately 1000 times as sensitive as anything we presently have.

While much of the effort over the next few years will be directed toward space, ground-based telescopes are not being neglected. Work began in 1985 on a project called the Very Long Baseline Array (VLBA). It should be completed in the mid-1990s, and will consist of 10 automated 25-meter dishes spaced across the continental United States, Hawaii, and St. Croix. The array will be controlled from the VLA in New Mexico.

The VLBA will be able to probe the radio waves emanating from deep within the nuclei of exploding galaxies; it will give us information from the region near the gigantic black hole that lies at the center. Motion pictures of jets and explosions may even be possible.

The Europeans also have a similar project underway (called VLN). Furthermore, the Australians have a project called AT which will include six 72-foot dishes at one point of Australia coupled with dishes in other regions.

It is also likely there will eventually be a link between these three systems, giving a radio telescope with an effective aperture the size of Earth. And plans are now underway to launch a satellite-borne radio telescope called Quasat which will be used in conjunction with this system.

Finally there are also plans for much larger ground-based optical telescopes. The National Optical Astronomical Observatory hopes to build a multimirror telescope similar to the one that is in operation at Mt. Hopkins in Arizona (which has an equivalent diameter of roughly 14 feet). It will use four 26-foot mirrors and will have an equivalent diameter of 52 feet. A project already underway on Mauna Kea, Hawaii (co-sponsored by

Caltech and the University of California) is referred to as the Keck telescope. It will have a segmented mirror (consisting of many fitted segments) roughly 32 feet across. Each segment will be independently controlled by computer. And in South America the European Southern Observatory is planning an array of four telescopes that will function as a unit. Each mirror will be 26 feet across.

Overall, the future of astronomy looks bright, and there is reason for optimism. Some of the deepest, most complex questions about galaxies, quasars, and the structure and origin of the universe may be answered within our lifetime.

Glossary

Absorption line   A dark line in a spectrum.

Accretion disk   A rotating disk formed by material attracted by a gravitating object.

Active galaxy   A galaxy that is emitting large amounts of energy from near its core.

Antennae   Refers to two galaxies that are colliding. Long antennae, or streams of gas emanate from the system.

Aperture   The diameter of the disk of a radio telescope, or diameter of mirror of optical telescope.

Asteroid   A small rocky object in orbit around the sun. Varies in size from a grain of sand to several hundred miles across.

Axion   A very light particle that is predicted by Grand Unified Theory.

Balmer series   Emission or absorption spectral lines in the spectrum of hydrogen.

Big bang theory   A cosmological theory that assumes the universe began as an explosion at a single point.

Binary system   A system of two stars, or a system of two galaxies.

Black hole   An object whose gravity is so strong that its escape velocity exceeds the velocity of light.

Bottom-up theory   A theory that assumes that galaxies formed first, clusters and superclusters later.

Cannibalism   The "eating" of one galaxy by another.

CCD   Charge-coupled device. A device for enhancing images.

Celestial mechanics A study of the motions of celestial objects.

Cepheid variable A star that changes in brightness periodically.

Closed universe A universe in which the recession of the galaxies eventually stops. Positively curved.

Cluster A group of stars, or galaxies.

Cold dark matter Dark matter of low velocity (e.g., axion).

Computer simulation An attempt to duplicate a process using a computer.

Constellation A group of stars that appear to be close to one another in the sky.

Cosmic background radiation Radiation that was released in the early universe. It now fills universe uniformly and has a temperature of 3 K.

Cosmic jet A narrow jet that emanates from the core of a galaxy.

Cosmic rays High-speed particles from space that strike our atmosphere, producing showers of other particles and radiation.

Cosmic string Hypothetical string in the early universe. May be responsible for structure.

Cosmology Study of the structure and evolution of the universe.

Critical density Density at which the universe is flat. Dividing line between open and closed universes.

Cutoff Refers to point beyond which there are no galaxies or quasars.

Dark matter Matter we are certain exists but cannot see.

Decoupling Refers to the release of radiation from matter in the early universe.

Density wave A wave of matter of varying density.

Differential rotation Rotational motion in which material or objects (e.g., stars in a galaxy) are at different distances from the center.

Disk galaxy Galaxy that has the shape of a disk. Spiral galaxies are disk galaxies.

Doppler effect A change in wavelength that occurs when the body emitting the waves is either approaching or receding from the observer.

Dwarf elliptical   A small elliptical galaxy.

Dwarf irregular   A small galaxy that has no form.

Dynamo   A machine for generating electricity.

Electromagnetic spectrum   The spectrum of radiations of various wavelengths, ranging from radio to gamma rays.

Elementary particle   A basic particle of nature that has no substructure.

Emission line   A bright spectral line.

Event horizon   Surface of a black hole. If you pass through it there is no escape.

False vacuum   An energy state of the early universe. Unlike a true vacuum, it has mass.

Faraday rotation   A rotation of the orientation of the polarization direction caused by magnetic fields.

Filaments   Long luminous strands of gas and stars that emanate from a galaxy.

Focal length   Distance from lens to focus when rays enter the lens in parallel.

Fluctuations   Changes in density from point to point.

Gamma ray   An energetic form of radiation.

Galactic dynamo   An "engine" at the center of a galaxy that generates energy.

General relativity   A theory devised by Einstein in 1915 to explain gravity.

Globular cluster   A roughly spherical array of from a few hundred thousand to a few million stars.

Great Attractor   A huge supercluster or "wall of clusters" that is attracting our Local Supercluster.

Hot dark matter   Dark matter that has high velocity.

Hubble flow   The rate at which galaxies are separating due to the expansion of the universe.

Hubble plot   A plot of redshift of galaxies versus their distance.

Hubble telescope   A large (94-inch) reflecting telescope that was put in space in 1990.

Inflation   Refers to a sudden change in the rate of expansion of the universe. Occurred shortly after the big bang.

Infrared   A region of the electromagnetic spectrum just beyond the visible region.

Interacting galaxies   Galaxies that are feeling one another's gravitational pull. Filaments are frequently pulled out.

Interferometer   Refers to an array of radio telescopes.

Interstellar gas   Gas between the stars.

Interstellar medium   A thin gas that permeates our galaxy.

Light-year   The distance light travels in one year.

Local Group   The group of galaxies that includes the Milky Way.

Local Supercluster   The group of clusters that includes the Local Group.

Merging galaxies   Galaxies that come together or merge into a single galaxy.

Messier object   Refers to object in a table of "fuzzy" celestial objects compiled by Charles Messier in 1787.

Meteor   A flash of light in the sky. Caused by a grain of sand striking our atmosphere at high speed.

Nebula   A cloud of interstellar gas and dust.

Neutral hydrogen   Hydrogen that contains both a proton and an electron. It has no net charge.

Neutron star   A star made up of neutrons. Usually only a few miles across.

N galaxy   A galaxy with an extremely dense nucleus.

Nova   A star that suddenly increases in brightness, then decreases slowly to its original brightness.

Nuclear bulge   Refers to the central bulge of a galaxy.

Nuclear fission   The breaking up, or fissioning, of the nucleus of an atom.

Nuclear reactor   A device that supplies nuclear energy.

Nucleus   The massive central portion of an atom. In relation to a galaxy, it is the central portion.

Neutrino   A high speed, elusive particle that has little or no mass.

Occultation   Eclipse.

Open universe   A universe that expands forever. Negatively curved.

Optical emission   Radiation that is emitted in the visible part of the electromagnetic spectrum.

Peculiar galaxy   A galaxy that appears strange. May have streamers coming out of it.

Peculiar velocity   Velocity in excess of Hubble flow.

Period-luminosity relation   The relation between the period of a Cepheid variable and its brightness.

Photometer   A device for measuring light intensity.

Planck time   Occurs 10^{-43} seconds after the big bang. At earlier times the universe is in the quantum realm.

Plasma   Ionized gas.

Polarization   Process in which electromagnetic waves with planes of oscillation in all but one direction are removed from a light beam.

Polar-ring galaxy   Galaxy with a ring around it that encircles the poles.

Primordial black hole   Black hole that is created in the early universe.

Quantum fluctuations   Fluctuations or changes in density that occur because of the uncertainty principle.

Quantum realm   Realm of atoms and elementary particles. Region where quantum theory is needed to describe processes.

Quantum theory   The branch of physics that deals with the structure and behavior of atoms and elementary particlesand their interaction with radiation.

Quasar   Energetic object in the outer regions of the universe.

Radiation   Photons. Electromagnetic energy.

Radio galaxy   Galaxy that emits radio waves.

Radio lobes   Two lobes on either side of a galaxy composed of plasma that emits radio waves.

Radio telescope   A device for detecting radio waves from celestial radio sources.

Redshift   A shift of spectral lines toward the red end of the spectrum. Indicates that the object is receding.

Relativity theory   Theory devised by Einstein to explain motion and gravity.

Ring galaxy   Galaxy that has a ring around it.

Rotation curve   A plot of the rotational speed of stars versus their distance from the center of the galaxy.

Schmidt telescope   A telescope that uses both lenses and a mirror.

Seyfert   A compact radio galaxy. More energetic than a usual radio galaxy.

Spectral line   A line that occurs when light is passed through a spectroscope.

Spectrogram   Photograph of a spectrum.

Spectroscopy   Study of spectral lines.

Starburst   The sudden generation of a large number of new stars near the center of a galaxy.

Sunspot   Region on the surface of the sun that is slightly cooler than surrounding regions. Appears dark.

Supercluster   A cluster of clusters.

Supernova   An exploding star.

Supersymmetric particle   Particle predicted by supersymmetry or supergravity theory.

Synchrotron radiation   The radiation emitted by charged particles moving through a magnetic field.

Thermal radiation   Radiation produced as a result of heat.

Tidal forces   Forces due to differences in gravitational pull.

Top-down theory   Theory that assumes that superclusters formed first, clusters later, then galaxies.

Uncertainty principle   A basic principle of quantum theory. Describes a "fuzziness" associated with nature on atomic scale.

Void   A region of space where there are no galaxies.

Wavelength   The distance between equivalent points on a wave.

Bibliography

The following is a list of general and technical references for the reader who wishes to learn more about the subject. References marked with an asterisk are of a more technical nature.

CHAPTER 1: Introduction

Editors of Time-Life Books, *Galaxies* (Alexandria, Virginia: Time-Life Books, 1989).
Ferris, Timothy, *Galaxies* (New York: Stewart, Tabori and Chang, 1982).
Kaufmann, William, III, *Galaxies and Quasars* (San Francisco: Freeman, 1979).

CHAPTER 2: Galaxies

Bartusiak, Marcia, *Thursday's Universe* (New York: Times Books, 1986).
Berendzen, Richard; Hart, Richard; and Seeley, Daniel, *Man Discovers the Galaxies* (New York: Science History Publications, 1978).
Ferris, Timothy, *Galaxies* (New York: Stewart, Tabori and Chang, 1982).
Ferris, Timothy, *The Red Limit* (New York: Morrow, 1977).
Harrison, Edward, *Cosmology* (London: Cambridge University Press, 1981).
Hodge, Paul, *Galaxies* (Cambridge: Harvard University Press, 1986).
Kaufmann, William, III, *Galaxies and Quasars* (San Francisco: Freeman, 1979).

Parker, Barry, "The Discovery of the Expanding Universe." *Sky and Telescope* (September 1986) 227.
Shapley, Harlow, *Through Rugged Ways to the Stars* (New York: Scribner's, 1969).
Silk, Joseph, *The Big Bang* (San Francisco: Freeman, 1980).
Whitney, Charles, *The Discovery of Our Galaxy* (New York: Knopf, 1971).

CHAPTER 3: The Discovery of Radio Sources

Hey, J. S., *The Evolution of Radio Astronomy* (New York: Neal Watson Publications, 1973).
Hey, J. S., *The Radio Universe* (New York: Pergamon Press, 1983).
Kellerman, K., and Sheets, B., *Serendipitous Discoveries in Radio Astronomy* (Green Bank, Virginia: National Radio Astronomy Observatory, 1983).
Lovell, Bernard, *The Story of Jodrell Bank* (New York: Harper and Row, 1984).
Sullivan, W. T., III, "A New Look at Karl Jansky's Original Data." *Sky and Telescope* (August 1978) 101.

CHAPTER 4: Exploding and Peculiar Galaxies

Downs, Ann, "Radio Galaxies." *Mercury* (March–April 1986) 66.
Goldsmith, Donald, "Exploding Galaxies." *Mercury* (January–February 1977) 2.
Kaufmann, William, III, *Galaxies and Quasars* (San Francisco: Freeman, 1979).
Kaufmann, William, III, "Exploding Galaxies and Supermassive Black Holes." *Mercury* (September–October 1978) 97.
Verschuur, Gerrit, *The Invisible Universe Revealed* (New York: Springer-Verlag, 1987).

CHAPTER 5: Cosmic Jets and Galactic Dynamos

Bartusiak, Marcia, *Thursday's Universe* (New York: Times Books, 1986).
Bladford, Roger; Begelman, Mitchell; and Rees, Martin, "Cosmic Jets." *Scientific American* (May 1982) 124.

Kanipe, Jeff, "M-87: Describing the Indescribable." *Astronomy* (May 1987) 6.

Rodriguez, Luis, "Cosmic Jets: Bipolar Outflows in the Universe." *Astronomy* (June 1984) 66.

Verschuur, Gerrit, *The Invisible Universe Revealed* (New York: Springer-Verlag, 1987).

CHAPTER 6: Quasars

Balik, B., "Quasars with Fuzz." *Mercury* (May–June 1983) 81.

Kaufmann, William, III, *Galaxies and Quasars* (San Francisco: Freeman, 1979).

Osmer, Patrick, "Quasars as Probes of the Distant and Early Universe." *Scientific American* (February 1982) 126.

Preston, Richard, *First Light: Search for the Edge of the Universe* (New York: New American Library, 1988).

*Schmidt, Martin, "3C 273: A Star-like Object with Large Redshift." *Nature* (March 16, 1963) 1040.

Shipman, Harry, *Black Holes, Quasars and the Universe* (Boston: Houghton-Mifflin, 1980).

*Wyckoff, Susan, "Resolvability of Quasar Images." *Astrophysical Journal* (August 1981) 750.

Wyckoff, Susan, and Wehinger, Peter, "Are Quasars Luminous Nuclei of Galaxies?" *Sky and Telescope* (March 1981) 200.

CHAPTER 7: Is Our Galaxy Exploding?

Chaisson, Eric, "Journey to the Center of the Galaxy." *Astronomy* (August 1980) 6.

Couper, Heather, "Journey to the Center of the Galaxy." *New Scientist* (April 26, 1984) 32.

Geballe, Thomas, "The Central Parsec of the Galaxy." *Scientific American* (July 1979) 60.

Parker, Barry, "Celestial Pinwheels: The Spiral Galaxies." *Astronomy* (May 1985) 14.

Waldrop, M., "Core of the Milky Way." *Science* (October 1985) 230.

CHAPTER 8: A Detailed Look at a Nearby Exploding Galaxy: Centaurus A

Burns, Jack, and Price, Marcus, "Centaurus A: The Nearest Active Galaxy." *Scientific American* (November 1983) 56.

Feigelson, Eric, and Schreier, E. J., "The X-ray Jets of Centaurus A and M87." *Sky and Telescope* (January 1983) 6.

Morrison, Nancy, and Gregory, S., "Centaurus A: The Nearest Active Galaxy." *Mercury* (May–June 1984) 75.

Verschuur, Gerrit, *The Invisible Universe Revealed* (New York: Springer-Verlag, 1987).

CHAPTER 9: Colliding Galaxies: The Discovery

Keel, William, "Crashing Galaxies: Cosmic Fireworks." *Sky and Telescope* (January 1989) 18.

Hartley, Karen, "Mixing It Up in Space." *Science News* (April 8, 1989) 207.

Toomre, Alar, and Toomre, Juri, "Violent Tides Between Galaxies." *Scientific American* (December 1973) 38.

*Toomre, Alar, and Toomre, Juri, "Galactic Bridges and Tails." *Astrophysical Journal* (December 1972) 623.

CHAPTER 10: Merging Galaxies

Barnes, Joshua, "Evolution of Compact Groups and the Formation of Elliptical Galaxies." *Nature* (March 9, 1989) 123.

Hartley, Karen, "Elliptical Galaxies Forged by Collisions." *Astronomy* (May 1985) 5.

Keel, William, "Binary and Multiple Galaxies and Their Interactions." *Reference Encyclopedia of Astronomy and Astrophysics* (New York: Van Nostrand, 1990).

Ostriker, J.P., "Elliptical Galaxies Are Not Made by Merging Spiral Galaxies." *Comments on Astrophysics* (January 1980) 227.

Schweizer, F., "Colliding and Merging Galaxies." *Science* (January 1986) 227.

Schweizer, F., "Merging Groups of Galaxies." *Nature* (March 9, 1989) 119.

CHAPTER 11: Collisions and Computers

*Barnes, Joshua, "Encounters of Disk/Halo Galaxies." *Astrophysical Journal* (August 1988) 699.
*Borne, Kirk, "Interacting Binary Galaxies: Matching Models and Observations." *Astrophysical Journal* (July 1988) 38.
*Keel, William; Kennicutt, Robert; Hummel, Ko; and van der Hulst, Thijs, "The Effects of Interactions on Spiral Galaxies: Nuclear Activity and Star Formation." *Astrophysical Journal* (May 1985) 708.
*Negroponte, John, and White, Simon, "Simulations of Mergers Between Disk/Halo Galaxies." *Monthly Notices of the Royal Astronomical Society* (January 1983) 205.

CHAPTER 12: Clusters and Superclusters

Gregory, Stephen, and Thompson, Laird, "Superclusters and Voids in the Distribution of Galaxies." *Scientific American* (March 1982) 106.
Helfand, David, "Superclusters and Large-Scale Structure of the Universe." *Physics Today* (October 1983) 17.
Marshall, Laurence, "Superclusters: Giants of the Cosmos." *Astronomy* (April 1984) 6.
Peebles, J., "The Origin of Galaxies and Clusters of Galaxies." *Science* (June 1984) 1385.
*Tully, Brent, "The Local Supercluster." *Astrophysical Journal* (June 1982) 389.
Waldrop, M., "The Large-Scale Structure of the Universe." *Science* (March 4, 1983) 1050.

CHAPTER 13: To the Ends of the Universe

Cornell, James, *Bubbles, Voids and Bumps in Time: The New Cosmology* (London: Cambridge University Press, 1989).

Gregory, Stephen, "The Structure of the Visible Universe." *Astronomy* (April 1988) 4.

Overbye, Dennis, "Exploring the Edge of the Universe." *Discover* (December 1982) 22.

Parker, Barry, *Creation* (New York: Plenum Press, 1988).

Scherrer, Robert, "Part One: From the Cradle of Creation." *Astronomy* (February 1988) 40.

Index